Better Homes and Gardens®

THE BEST OF WOOD

BOOK 3

WE CARE!

All of us at Meredith® Books are dedicated to giving you the
information and ideas you need to create beautiful and useful
woodworking projects. We guarantee your satisfaction with this book
for as long as you own it. We also welcome your comments and
suggestions. Please write us at Meredith® Books, RW-240,
1716 Locust St., Des Moines, IA 50309-3023.

© Copyright 1995 by Meredith Corporation, Des Moines, Iowa. All Rights Reserved. Printed in the United States of America.
First Edition. Printing Number and Year: 5 4 3 2   99 98 97 96 95
ISSN: 1069-5028. ISBN: 0-696-02583-3.

WOOD® Books
An Imprint of Meredith® Books

THE BEST OF WOOD BOOK 3
Editor: Benjamin Allen
Associate Art Director: Tom Wegner
Production Manager: Douglas Johnston

Vice President and Editorial Director: Elizabeth P. Rice
Executive Editor: Kay Sanders
Art Director: Ernest Shelton
Managing Editor: Christopher Cavanaugh

President Book Group: Joseph J. Ward
Vice President, Retail Marketing: Jamie L. Martin
Vice President, Direct Marketing: Timothy Jarrell

On the front cover: Picnic-Perfect Table and Benches, pages 38–41
On the back cover: Sunny-Day Sandbox, pages 76–79; Signature Shop Clock, pages 154–159; Tour-De-Force Tureen, pages 126–130; Do-It-All Mobile Tablesaw Base, pages 7–11

Meredith Corporation
Chairman of the Executive Committee: E.T. Meredith III
Chairman of the Board and Chief Executive Officer: Jack D. Rehm
President and Chief Operating Officer: William T. Kerr

Special thanks to Larry Clayton, Editor of WOOD magazine; Marlen Kemmet, Senior Editor/How-To of WOOD magazine; Larry Johnston, Special-Interest Editor of WOOD magazine; and the WOOD magazine staff.

Year after year WOOD® Magazine consistently publishes a wide variety of beautiful projects for every skill level of woodworker. We invite you to look back on our best projects of 1994. Because every project is designed, built, and checked to our exacting standards, we know you'll get results you'll be proud of for years.

Contents

Small Gifts 101

Decorating Accessories 131

Acknowledgements 160

Workshop Wonders

Building accessories to make your shop a more efficient space is not only practical and economical, but it continues the proud woodworker tradition of self-sufficiency. Every workshop project in The Best of WOOD Book 3 comes from our Idea Shops where we rigorously test them, so you can rest assured they are the best designs for shop projects anywhere.

Do-It-All Mobile Tablesaw Base

It's sad but true that the space beneath most tablesaws goes to waste. But that needn't be the case, as you can see here. Our cabinet stores a plentiful supply of sawblades, router bits, and other woodworking gear on the sliding trays. Plus, the entire cabinet is easy to move around in your shop, thanks to a pair of casters. Just store it against the wall when the cutting is done.

Note: Our cabinet was made to fit a Ryobi BT3000 benchtop tablesaw. You might need to change the overall dimensions to fit your particular benchtop saw.

START WITH THE PLYWOOD PANELS AND EDGING

1 Cut the sides (A, B) to the sizes listed in the Bill of Materials from ¾" birch plywood. Note that side A is ¾" shorter than side B to allow for

clearance against the floor later after adding the casters.

2 Cut the back (C) and top and bottom (D) to size.

3 From the edge of ¾"-thick stock, rip the maple edging (E, F, G, H) to size plus 1" in length. Glue and clamp the edging to the edges of the plywood panels where shown on the Exploded View drawing. Later, trim the ends of the edging flush.

continued

Do-It-All Mobile Tablesaw Base

continued

4 Mark the locations of the rabbets, grooves, and dadoes in the plywood panels (A, B, C), and cut or rout them to the sizes listed on the Exploded View drawing. (We fit our tablesaw with a ¾" dado blade and wooden auxiliary rip fence to cut the dadoes, rabbets, and grooves. Then, we test-cut scrap pieces first to verify a snug fit of the plywood pieces in the dadoes and grooves.)

ADD THE TRAY SUPPORTS

1 From ¼" hardboard, cut the tray supports (I, J) to size.

2 Position and screw the bottom supports (J) 1" above the bottom dado. Then, using a piece of ¾"-thick stock as a spacer, position the remaining supports (I), drill screw-mounting holes, and screw the supports in place as shown in the photo *above right*. (Note that we placed a long piece of ¾"-thick walnut in the bottom ¾" dadoes to align the two side panels (A, B) before attaching the supports.)

LET THE ASSEMBLY BEGIN

1 Dry-clamp the plywood panels (A,B,C D) together to check the fit, and trim if necessary. Then, glue and clamp the pieces, checking for square. (To prevent unsightly glue stains on the plywood, we placed masking tape next to the glue joints before gluing. Later, after the glue dried, we peeled off the tape, taking the glue squeeze-out with it.)

2 Measure the length of the opening, and cut the toekick (K) to shape, tapering the corner where dimensioned on the Wheel detail accompanying the Exploded View drawing. Drill countersunk screw holes, and glue and screw the toekick in place.

3 Turn the assembled tablesaw base upside down, drill the mounting holes, and screw the 2½" rigid casters in place.

4 Cut the border strips (L) to size. Drill the countersunk screw holes where shown on the Exploded View drawing. Screw the strips to the top of the cabinet.

ADD THE THREE TRAYS FOR PLENTY OF STORAGE

1 From ¾" maple, rip and crosscut the tray fronts (M, N) to size. From ¾" plywood, cut the tray bottoms (O) to size.

2 Transfer the full-sized handle-cutout pattern *below* to poster board. Cut the poster board template to shape. Center the template on the front face of each tray front. Trace its outline onto each tray front, and cut the radius to shape to create the concave opening. Now, sand the tray fronts smooth.

3 Cut a ¾" rabbet ⅜" deep along the bottom back edge of each tray front (M, N).

4 Cut a cleat (P) to size for each tray. Drill six countersunk holes in each cleat. With the bottom edge
continued

For evenly spaced tray supports, use a piece of ¾"-thick stock to position the pieces before screwing them in place.

Bill of Materials

Part	Finished Size*			Mat.	Qty.
	T	W	L		
A side	¾"	19½"	19¼"	BP	1
B side	¾"	19½"	20"	BP	1
C back	¾"	23¼"	20"	BP	1
D top & bottom	¾"	23¼"	19⅜"	BP	2
E edging	¼"	¾"	19¼"	M	2
F edging	¼"	¾"	20"	M	2
G edging	¼"	¾"	20"	M	2
H edging	¼"	¾"	23¼"	M	3
I supports	¼"	3¾"	18½"	HB	4
J supports	¼"	5½"	18½"	HB	2
K toekick	¾"	3¼"	22½"	M	1
L strips	¾"	1½"	20"	M	2
M fronts	¾"	4⅜"	22⁷⁄₁₆"	M	2
N front	¾"	6⅛"	22⁷⁄₁₆"	M	1
O bottoms	¾"	18¼"	22⁷⁄₁₆"	BP	3
P cleats	¾"	1"	21⁷⁄₁₆"	M	3
Q holders	1½	2⁷⁄₁₆"	21⁷⁄₁₆"	F	8
R slides	¼"	1"	18½"	HB	2
S holders	¾"	2¼"	3½"	M	2

Length measured with the grain.

Materials Key: BP–birch plywood, M–maple, HB–hardboard, F–fir.

Supplies: #8x¾", #8x1¼", #8x1½", #8x2¼", #8x2¾" flathead wood screws, 2–2³⁄₁₆" (overall length) screw eyes, ¼x1x15" aluminum bar stock for locking bar, 2–2½" rigid casters (3¼" overall height) with 8–#8x¾" panhead sheet-metal screws, ⅜" dowel stock, ½" dowel stock, primer, red enamel paint, clear finish.

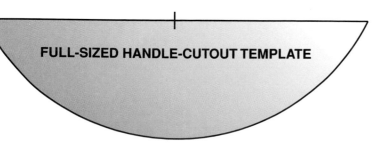

FULL-SIZED HANDLE-CUTOUT TEMPLATE

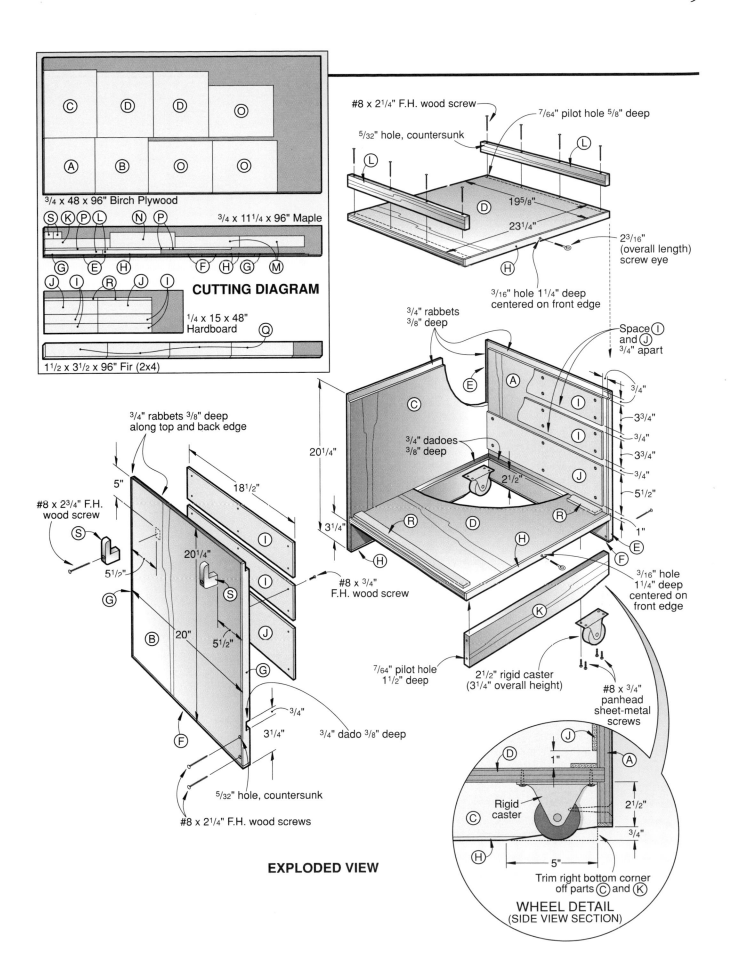

CUTTING DIAGRAM

3/4 x 48 x 96" Birch Plywood

3/4 x 11 1/4 x 96" Maple

1/4 x 15 x 48" Hardboard

1 1/2 x 3 1/2 x 96" Fir (2x4)

#8 x 2 1/4" F.H. wood screw

7/64" pilot hole 5/8" deep

5/32" hole, countersunk

19 5/8"

23 1/4"

2 3/16" (overall length) screw eye

3/16" hole 1 1/4" deep centered on front edge

3/4" rabbets 3/8" deep

Space I and J 3/4" apart

3/4"

3 3/4"

3/4"

3 3/4"

3/4"

5 1/2"

1"

3/4" dadoes 3/8" deep

2 1/2"

3/16" hole 1 1/4" deep centered on front edge

3/4" rabbets 3/8" deep along top and back edge

5"

20 1/4"

18 1/2"

#8 x 2 3/4" F.H. wood screw

20 1/4"

#8 x 3/4" F.H. wood screw

20 1/4"

3 1/4"

5 1/2"

20"

5 1/2"

3/4"

3 1/4"

3/4" dado 3/8" deep

5/32" hole, countersunk

#8 x 2 1/4" F.H. wood screws

EXPLODED VIEW

7/64" pilot hole 1 1/2" deep

2 1/2" rigid caster (3 1/4" overall height)

#8 x 3/4" panhead sheet-metal screws

1"

Rigid caster

2 1/2"

3/4"

5"

Trim right bottom corner off parts C and K

WHEEL DETAIL
(SIDE VIEW SECTION)

Do-It-All Mobile Tablesaw Base

continued

of the cleat flush with the top edge of the rabbet, glue and screw the cleats to the tray fronts. Glue and screw a tray front/cleat assembly to each tray bottom (O), checking that the front is square to the tray bottom.

5 To make the router bit holders (Q), cut four pieces of 2×4 stock to 21⁷⁄₁₆" long. Mark a centerline along *both edges* (not surfaces) of each piece of 2×4 stock. Starting 1²³⁄₃₂" from the ends and spacing the centerpoints 2" apart, mark centerpoints on the 2×4".

6 Drill ¼" and ½" router-bit shank holes in the 2×4 stock. The number of ¼" holes vs. ½" holes will be determined by your bit collection. (After drilling into the wood, we wobbled the bit slightly to allow the bit shanks to be removed easily from the holes.)

7 Rout ¼" round-overs along all edges of each piece of 2×4.

8 Using the Bit Holder detail on the *opposite page* for reference, angle the blade on your tablesaw, and rip each 2×4 section in two. Sand each holder (Q) smooth.

9 Locate and drill countersunk screw holes on the bottom side of each tray (O). Screw the bit holders to the top of each tray.

10 Use the Bottom Tray drawing to assemble the blade, shim, and dado-cutter holders.

11 Cut the drawer slides (R) to size, and glue and clamp them to the cabinet bottom (D), flush against the cabinet sides (A, B).

CONSIDER A FEW ADD-ONS

1 For hanging the miter-gauge extension on the side of the cabinet, twice transfer the miter-gauge extension holder full-sized pattern to ¾"-thick stock. Cut the holders (S) to shape, drill a screw hole in each, and screw the holders to the side of the cabinet where shown on the Exploded View drawing.

2 For added security, drill pilot holes and add a pair of 2¹³⁄₁₆" screw eyes to the cabinet where shown on the Exploded View drawing. Then, crosscut a piece of ¼×1" aluminum bar stock to 15" long for the locking bar (see the Locking Bar drawing *below right* for reference). Drill a ⁵⁄₁₆" hole at each end to align with the screw-eye holes. Later, use padlocks to secure the locking bar to the screw eyes.

SAND IT, PAINT IT, AND BRING ON THE TABLESAW

1 Remove the rigid casters and screw eyes, and finish-sand the entire cabinet.

2 Apply two coats of clear finish to the trays.

3 Apply a coat of primer to the cabinet. Later, apply two coats of red enamel paint to the cabinet. When the paint has dried, reattach the screw eyes and casters.

4 Position the tablesaw on the base, drill mounting holes, and secure with bolts.

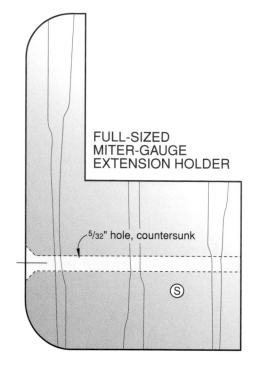

FULL-SIZED MITER-GAUGE EXTENSION HOLDER

⁵⁄₃₂" hole, countersunk

Ⓢ

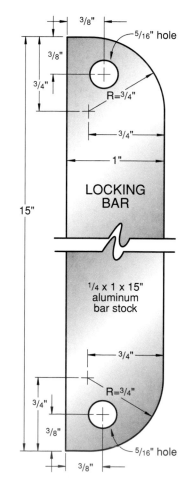

3/8"

5/16" hole

3/8"

3/4"

R=³⁄₄"

3/4"

1"

LOCKING BAR

15"

¼ × 1 × 15" aluminum bar stock

3/4"

R=³⁄₄"

3/4"

3/8"

5/16" hole

3/8"

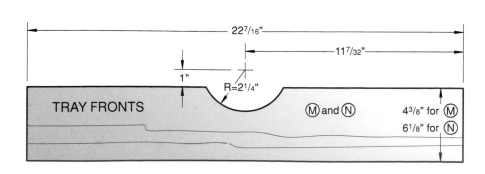

22⁷⁄₁₆"

11⁷⁄₃₂"

1"

R=2¼"

TRAY FRONTS

Ⓜ and Ⓝ

4³⁄₈" for Ⓜ

6¹⁄₈" for Ⓝ

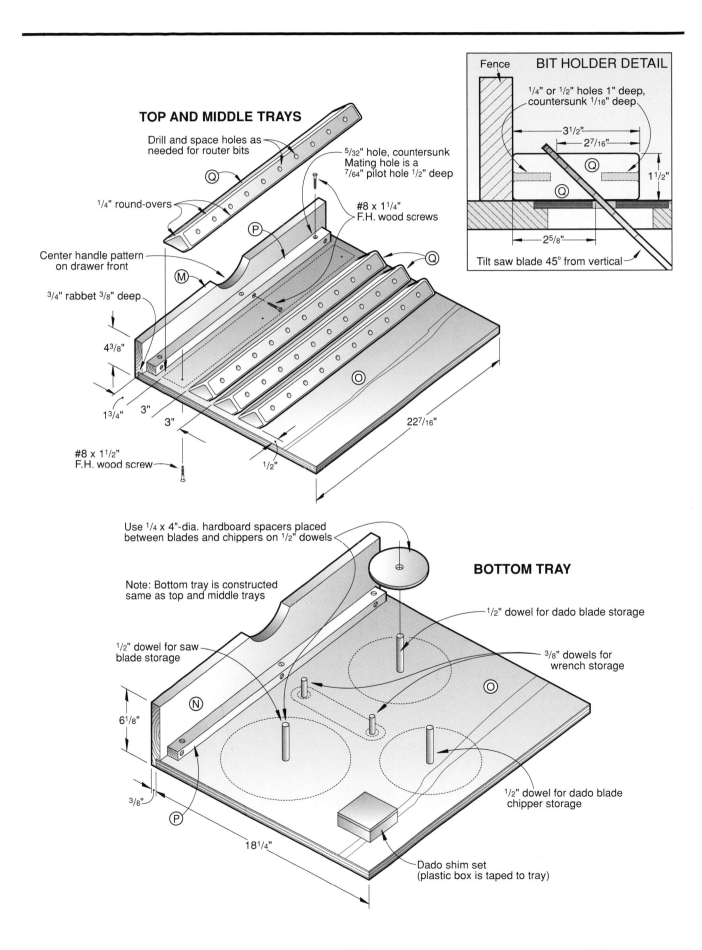

TOP AND MIDDLE TRAYS

Drill and space holes as
needed for router bits

1/4" round-overs

Center handle pattern
on drawer front

3/4" rabbet 3/8" deep

4 3/8"

1 3/4"

3"

3"

1/2"

22 7/16"

#8 x 1 1/2"
F.H. wood screw

5/32" hole, countersunk
Mating hole is a
7/64" pilot hole 1/2" deep

#8 x 1 1/4"
F.H. wood screws

BIT HOLDER DETAIL

Fence

1/4" or 1/2" holes 1" deep,
countersunk 1/16" deep

3 1/2"

2 7/16"

1 1/2"

2 5/8"

Tilt saw blade 45° from vertical

BOTTOM TRAY

Use 1/4 x 4"-dia. hardboard spacers placed
between blades and chippers on 1/2" dowels

Note: Bottom tray is constructed
same as top and middle trays

1/2" dowel for saw
blade storage

1/2" dowel for dado blade storage

3/8" dowels for
wrench storage

1/2" dowel for dado blade
chipper storage

6 1/8"

3/8"

18 1/4"

Dado shim set
(plastic box is taped to tray)

Accommodating Cabinets

Perforated hardboard and hooks help you get organized once and for all

What's the #1 complaint most woodworkers have about their shop? You guessed it. They have a hard time finding the right tool when they need it most. If your shop could stand a little bit more order, try building the double-door cabinet shown here or the single-door unit on *page 14*. You can clean up your act in a hurry, which will leave you more time to do what you really want to do— build stuff.

BEGIN WITH THE DOUBLE-DOOR CABINET AND DOOR FRAMES

1 Using the Cutting Diagram for reference, lay out and cut the cabinet sides (A), cabinet top and bottom (B), door sides (C), and door top and bottom (D) to the sizes listed in the Bill of Materials. We used a stop when cutting the pieces

to length to ensure that the height of the assembled doors would match that of the cabinet.

2 Set your tablesaw rip fence ¼" from the inside edge of the blade, and rip the edging strips (E, F, G, H) from the edge of ¾"-thick maple. Crosscut the edging strips to length plus 1". (We used the edging to hide the exposed plies of the plywood pieces.)

3 Glue and clamp the edging strips to the plywood pieces (A, B, C, D), and later trim the ends of the edging flush with that of the plywood.

4 As dimensioned on the Rabbet detail accompanying the Exploded View drawing, cut a ⅜" dado ⅜" deep ⅜" from the top and bottom ends of the cabinet and door sides A/F and C/H. Next, cut a ⅜"

continued

Bill of Materials

Part	Finished Size*			Mat.	Qty.
	T	W	L		
DOUBLE-DOOR CABINET					
A cab. sides	¾"	6¾"	48"	BP	2
B cab. top & bottom	¾"	6¾"	47¼"	BP	2
C dr. sides	¾"	2¼"	48"	BP	2
D dr. tops & bottoms	¾"	2¼"	23¼"	BP	4
E* cab. top & btm. edging	¼"	¾"	47¼"	M	2
F* cab. side edging	¼"	¾"	48"	M	2
G* dr. top & btm. edging	¼"	¾"	23¼"	M	4
H* dr. side edging	¼"	¾"	48"	M	4
I side cleats	¾"	1½"	43½"	P	6
J cab. top & btm. cleats	¾"	1½"	46½"	P	2
K dr. top & btm. cleats	¾"	1½"	22½"	P	4
L cab. pegbrd.	¼"	46½"	46½"	PB	1
M dr. pegbrd.	¼"	22½"	46½"	PB	2
N door panels	½"	22"	46"	BP	2
O* sides	½"	1"	48"	M	4
P* top & btm. bands	½"	1"	24"	M	4

*Initially cut parts marked with an * oversized. Then, trim each to finished size according to the how-to instructions.

Materials Key: BP–birch plywood, M–maple, P–pine, PB–pegboard.

Supplies: 4d finish nails, #12x¾" panhead sheet metal screws, 3" deck screws, 2–1½x48" continuous hinge, 2–3" wire pulls, 2½" hasp, 2" barrel bolt, padlock, enamel paint, satin polyurethane.

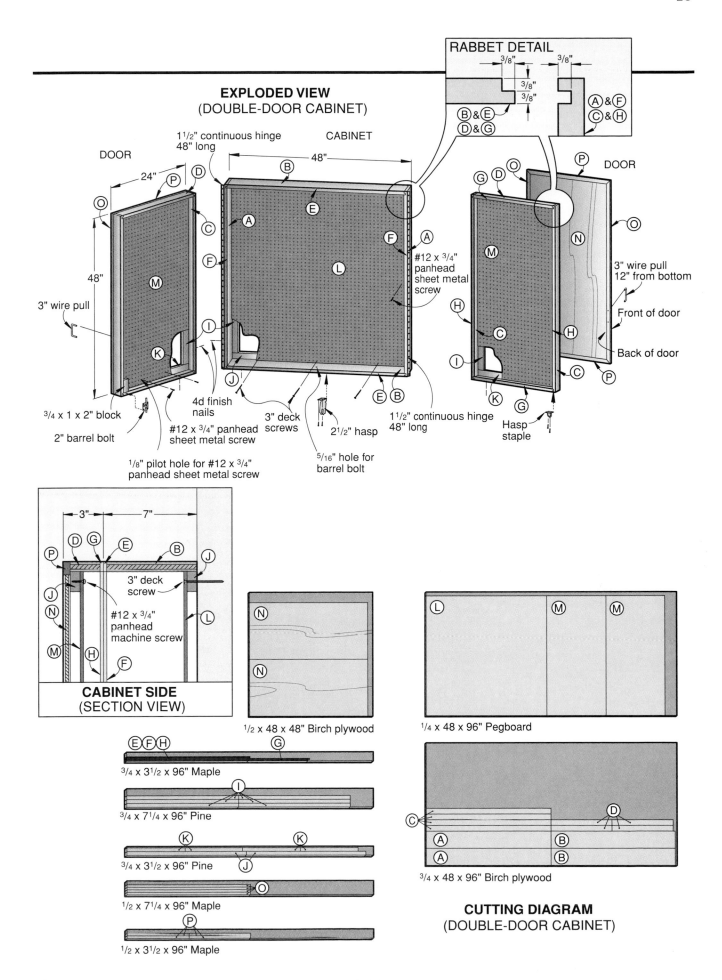

RABBET DETAIL

3/8"   3/8"
3/8"
3/8"

Ⓑ&Ⓔ
Ⓓ&Ⓖ

Ⓐ&Ⓕ
Ⓒ&Ⓗ

EXPLODED VIEW
(DOUBLE-DOOR CABINET)

1 1/2" continuous hinge
48" long

CABINET

DOOR

24"

48"

3" wire pull

48"

3/4 x 1 x 2" block

2" barrel bolt

#12 x 3/4" panhead
sheet metal screw

1/8" pilot hole for #12 x 3/4"
panhead sheet metal screw

4d finish
nails

3" deck
screws

2 1/2" hasp

5/16" hole for
barrel bolt

#12 x 3/4"
panhead
sheet metal
screw

1 1/2" continuous hinge
48" long

DOOR

3" wire pull
12" from bottom

Front of door

Back of door

Hasp
staple

CABINET SIDE
(SECTION VIEW)

3"   7"

3" deck
screw

#12 x 3/4"
panhead
machine screw

1/2 x 48 x 48" Birch plywood

1/4 x 48 x 96" Pegboard

Ⓔ Ⓕ Ⓗ   Ⓖ

3/4 x 3 1/2 x 96" Maple

Ⓘ

3/4 x 7 1/4 x 96" Pine

Ⓚ   Ⓚ

Ⓙ

3/4 x 3 1/2 x 96" Pine

Ⓞ

1/2 x 7 1/4 x 96" Maple

Ⓟ

1/2 x 3 1/2 x 96" Maple

3/4 x 48 x 96" Birch plywood

CUTTING DIAGRAM
(DOUBLE-DOOR CABINET)

Accommodating Cabinets

continued

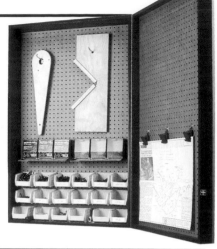

rabbet ⅜" deep along the ends of the top and bottom pieces B/E and D/G. (We test-cut scrap pieces of stock first to verify the settings.) Check the fit of the mating cabinet and door pieces.

5 Glue and clamp the cabinet assembly together, checking for square. Repeat the process to assemble the two doors.

NOW, ADD THE CLEATS AND PEGBOARD

1 Cut the cabinet and door cleats (I, J, K) to size from ¾" pine.

2 Glue and nail the cleats to the inside of the cabinet and doors flush with the back edge of the plywood panels where shown on the Exploded View and Side Section View drawings.

3 Measure the openings and cut the cabinet pegboard (L) and door pegboard (M) panels to size.

CUT THE DOOR PANELS AND BANDING TO SIZE

1 Cut the two plywood door panels (N) to size from ½" plywood.

2 Cut the door banding pieces (O, P) to size plus 1" in length from maple stock.

3 Rout a ¹/₁₆" chamfer along the front mating edges of the door panel (N) and banding (O, P). When joined together in the next step, the chamfers form a decorative V-groove. (See the Single Door Cabinet Exploded View drawing *opposite*.)

4 Miter-cut the banding pieces (O, P) to length. Glue and clamp them to the door panels (N).

5 With the edges and ends flush, glue and clamp the doors (I, J, K) to the door frames (C–H).

6 Lay the cabinet, front side facing up, on your workbench. Position

the doors, front side (or paneled side) up, on the cabinet with the edges and ends flush. To keep the hinges from binding later, lay a piece of folded writing paper between the cabinet and doors. The paper acts as a spacer to create the correct gap.

7 Use masking tape to secure the 1½×48" continuous (piano) hinges in place. Using the existing holes in the hinges as guides, drill ¹/₁₆" pilot holes into the cabinet and doors. Attach the hinges, and set the cabinet assembly upright, and open and close the doors to check the fit.

8 Locate and drill the holes, and attach a 3" wire pull to each door.

9 If locking the cabinet is a consideration, apply the barrel bolt, hasp, and hasp staple to the cabinet and doors.

FINISHING TOUCHES

1 Remove the hardware from the the assembled cabinet. Sand all the parts smooth.

2 Apply a clear finish (we used satin polyurethane) to the door banding O and P; see the intro photo for reference.

3 After the finish dries, mask off the edging, and apply primer to the rest of the surfaces. Later, paint the primed surfaces.

4 Fit the pegboard panels (L, M) in their respective openings. Use panhead sheet metal screws to secure the pegboard panels to the cabinet and doors.

5 Reattach the doors to the cabinet. With a helper, hang the cabinet on the wall, being sure to hit all available studs. Reattach the pulls, barrel bolt, and hasp.

HOW TO BUILD A SINGLE-DOOR CABINET

Using the Single-Door Cabinet Bill of Materials, Exploded View drawings, Cutting Diagram, and the same construction procedure as used on the double-door cabinet, build the single-door unit.

Bill of Materials					
Part	Finished Size*			Mat.	Qty.
	T	W	L		
SINGLE-DOOR CABINET					
A cab. sides	¾"	6¾"	48"	BP	2
B cab. top & bottom	¾"	6¾"	29¼"	BP	2
C dr. sides	¾"	2¼"	48"	BP	2
D dr. tops & bottoms	¾"	2¼"	29¼"	BP	2
E* cab. top & btm. edging	¼"	¾"	29¼"	M	4
F* cab. side edging	¼"	¾"	48"	M	4
G dr. top & btm. cleats	¾"	1½"	28½"	P	4
H dr. side cleats	¾"	1½"	43½"	P	4
I door panel	½"	28"	46"	BP	1
J* side bands	½"	1"	48"	M	2
K* top & btm. bands	½"	1"	30"	M	2
L dr. & cab pegboard	¼"	28½"	46½"	PB	2

*Initially cut parts marked with an * oversized. Then, trim each to finished size according to the how-to instructions

Materials Key: BP–birch plywood, P–pine, M–maple, PB–pegboard.

Supplies: 4d finish nails, 3" deck screws, #12x¾" panhead sheet metal screws, 1½x48" continuous (piano) hinge, 3" wire pull, 2½" hasp, lock, enamel paint, satin polyurethane.

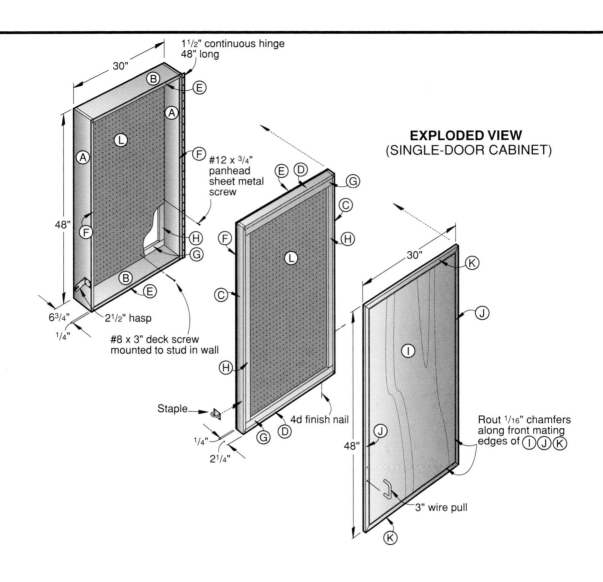

1¹/₂" continuous hinge 48" long

30"

48"

6³/₄"

1/4"

2¹/₂" hasp

#12 x ³/₄" panhead sheet metal screw

#8 x 3" deck screw mounted to stud in wall

Staple

1/4"

2¹/₄"

EXPLODED VIEW
(SINGLE-DOOR CABINET)

30"

48"

4d finish nail

Rout ¹/₁₆" chamfers along front mating edges of (I)(J)(K)

3" wire pull

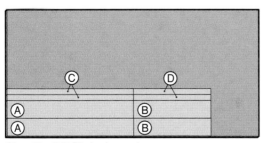

³/₄ x 48 x 96" Birch plywood

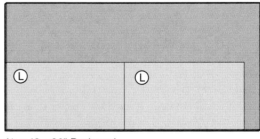

¹/₄ x 48 x 96" Pegboard

³/₄ x 7¹/₄ x 96" Pine

³/₄ x 3¹/₂ x 96" Maple

¹/₂ x 3¹/₂ x 96" Maple

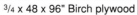
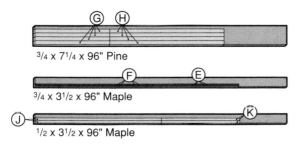

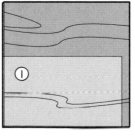

CUTTING DIAGRAM
(SINGLE-DOOR CABINET)

Drop-Leaf Mobile Workbench

Super-sturdy and loaded with storage

A rock-solid workbench that's mobile? In the past, this project might have sounded a little far-fetched, but not any more. A while back, WOOD® magazine reader Erv Roberts shared with us his garage/shop workbench design. Wanting to keep his cars inside during Iowa's cold, snowy winters, Erv needed a workbench with wheels so he could push it out of the way when the time came to bring his cars inside. He also wanted a workbench with fold-down leaves, giving him the most efficient against-the-wall storage possible. When Erv's hobby calls, his benchtop folds out to a full 36×58".

START WITH THE BASE CABINET FOR A SOLID FOUNDATION

1 Cut the two base side panels (A) to the size listed in the Bill of Materials from ¾" birch plywood.

2 Cut the base banding pieces (B, C) to size from ¾" maple. Glue the banding pieces to the panels where shown on the Base Side Panel drawing on the *opposite page*. Sand the banding flush.

3 Using the same drawing for reference, mark the location of, and cut or rout ¾" dadoes and grooves ¼" deep into the mating inside surface of each plywood side panel.

4 Mark and cut a ¾×5" taper on the bottom rear corner of each side panel. When assembling the base later, the tapers must be directly across from each other. The tapers allow the casters to come in contact with the floor without the corners of the side panels rubbing against the floor when moving the workbench.

5 Cut the shelves (D) and dividers (E, F) to size from ¾" plywood. Now, cut a pair of notches in the upper corners of the top divider (E) where shown on the Base Cabinet drawing.

6 Cut the banding strips (G), and band both ends of each shelf (D).

7 Glue and clamp the shelves and dividers between the side panels, checking for square.

8 Cut the toe kick (H) and cleats (I) to size. Drill mounting holes through the cleats, attach them, and then screw the toe kick in place so the outside face of the toe kick is recessed ¼" from the outside surface of banding pieces C and G.

ADD THE PULL-OUT HANDLES AND CASTERS

1 Cut the handles (J) to size. Joint or plane each to ¹¹⁄₁₆" thick. Transfer the full-sized Handle Pattern to one end of each handle blank. Cut the contoured ends to shape. Then, rout or sand ⅛" round-overs along all edges.

2 Drill the ½" and ⅜" holes in the handle where dimensioned on the Handle Pattern. Cut four ⅜" dowels to 1½" long and glue them into the holes. The dowels act as stops when the handles are pushed and pulled in and out of the cabinet.

3 Cut the handle retainer cleats (K, L) and retainers (M) to size. Glue and screw the bottom cleats (L) to the retainers (M).

4 Turn the base cabinet upside down, and position the rigid casters flush with the inside face of the side panels (A) and outside face of the shelf banding (G). Drill pilot holes and screw the casters in place.

continued

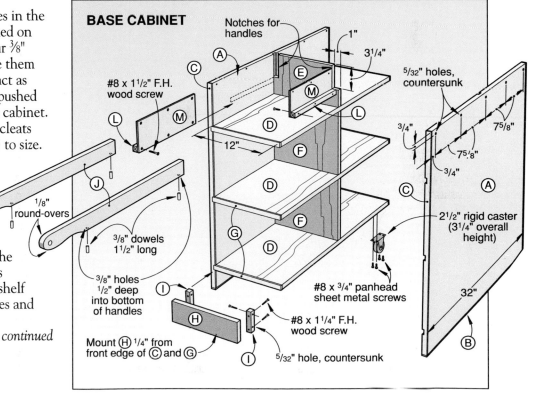

BASE SIDE PANEL

¼"
8½"
¾"
31½"
¾" groove ¼" deep
9¼"
¼"
32½"
9¼"
¾"
C   A   C
¾"
5"
¾" dadoes ¼" deep
B
Taper corner to allow rigid caster clearance
Note: Cut dadoes and groove on inside surface of each side panel
¾"
15⅝"   32"
3"
¼"
Note: Apply maple banding B and C to plywood before cutting dadoes and groove

BASE CABINET

Notches for handles
1"
A
3¼"
C
E
M
#8 x 1½" F.H. wood screw
⁵⁄₃₂" holes, countersunk
L
M
L
D
7⅝"
¾"
J
12"
F
7⅝"
¾"
D
A
⅛" round-overs
⅜" dowels 1½" long
G
F
2½" rigid caster (3¼" overall height)
⅜" holes ½" deep into bottom of handles
I
D
C
#8 x ¾" panhead sheet metal screws
H
I
32"
Mount H ¼" from front edge of C and G
#8 x 1¼" F.H. wood screw
⁵⁄₃₂" hole, countersunk
B

Drop-Leaf Mobile Workbench

continued

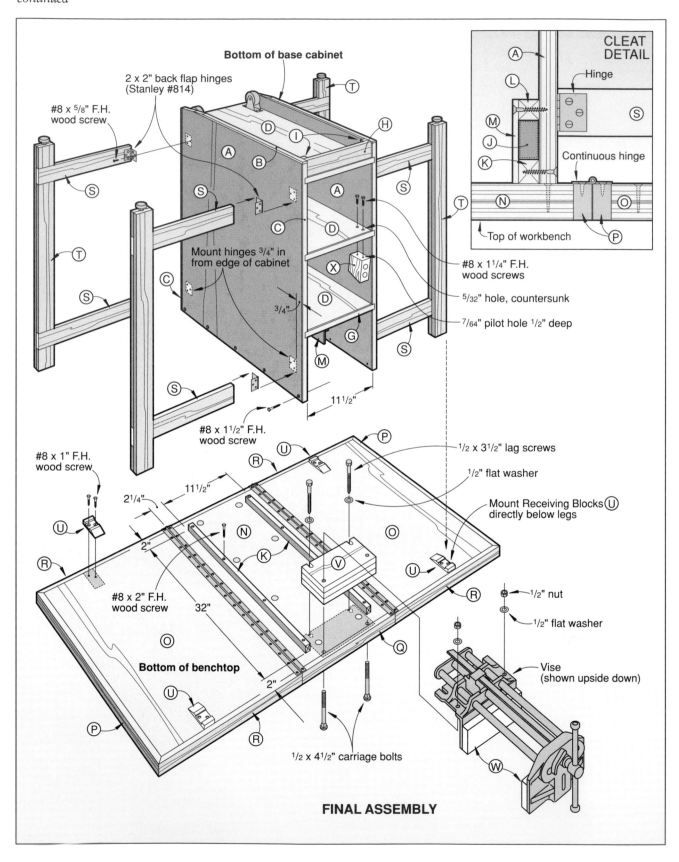

Bottom of base cabinet

2 x 2" back flap hinges
(Stanley #814)

#8 x ⁵⁄₈" F.H.
wood screw

Ⓐ Ⓑ Ⓒ Ⓓ Ⓘ Ⓗ Ⓣ Ⓢ Ⓧ Ⓜ Ⓖ

Mount hinges ³⁄₄" in
from edge of cabinet

³⁄₄"

11¹⁄₂"

#8 x 1¹⁄₂" F.H.
wood screw

**CLEAT
DETAIL**

Hinge

Ⓐ Ⓛ Ⓜ Ⓙ Ⓚ Ⓢ

Continuous hinge

Ⓝ Ⓞ Ⓟ

Top of workbench

#8 x 1¹⁄₄" F.H.
wood screws

⁵⁄₃₂" hole, countersunk

⁷⁄₆₄" pilot hole ¹⁄₂" deep

#8 x 1" F.H.
wood screw

2¹⁄₄"

11¹⁄₂"

2"

#8 x 2" F.H.
wood screw

32"

2"

Bottom of benchtop

Ⓤ Ⓡ Ⓝ Ⓚ Ⓞ Ⓥ Ⓠ Ⓟ

¹⁄₂ x 3¹⁄₂" lag screws

¹⁄₂" flat washer

Mount Receiving Blocks Ⓤ
directly below legs

¹⁄₂" nut

¹⁄₂" flat washer

Vise
(shown upside down)

Ⓦ

¹⁄₂ x 4¹⁄₂" carriage bolts

FINAL ASSEMBLY

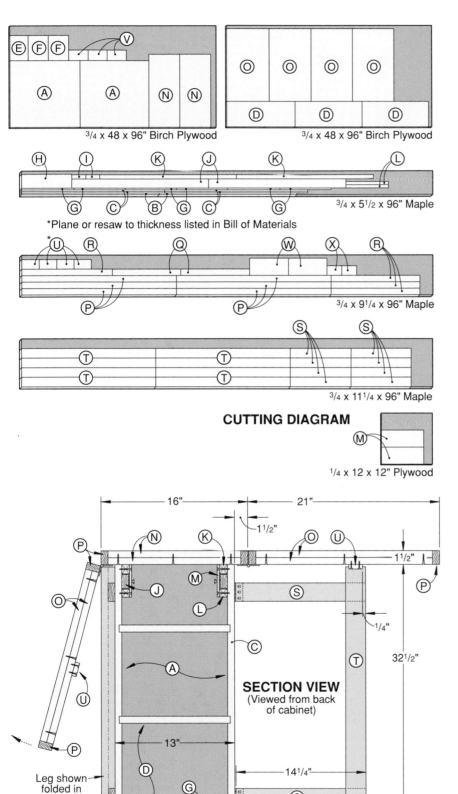

¾ x 48 x 96" Birch Plywood

¾ x 48 x 96" Birch Plywood

¾ x 5½ x 96" Maple

*Plane or resaw to thickness listed in Bill of Materials

¾ x 9¼ x 96" Maple

¾ x 11¼ x 96" Maple

CUTTING DIAGRAM

¼ x 12 x 12" Plywood

SECTION VIEW
(Viewed from back of cabinet)

16"
21"
1½"
1½"
¼"
32½"
13"
14¼"

Leg shown folded in

2½" rigid caster

Bill of Materials

Part	Finished Size*			Mat.	Qty.
	T	W	L		
BASE					
A sides	¾"	31½"	32¼"	BP	2
B bottom bands	¼"	¾"	32"	M	2
C side bands	¼"	¾"	32¼"	M	4
D shelves	¾"	12"	31½"	BP	3
E divider	¾"	12"	8½"	BP	1
F dividers	¾"	12"	9¼"	BP	2
G shelf bands	¼"	¾"	12"	M	6
H toe kick	¾"	3¼"	11½"	M	1
I cleats	¾"	1"	3¼"	M	2
HANDLE ASSEMBLY					
J handles	¹¹⁄₁₆"	2⅛"	32"	M	2
K cleats	¾"	1"	32"	M	2
L cleats	¾"	¾"	10"	M	2
M retainers	¼"	4"	10"	P	2
BENCHTOP					
N* center	¾"	34½"	14½"	BP	2
O* leaves	¾"	34½"	19½"	BP	4
P banding	¾"	1½"	36"	M	6
Q banding	¾"	1½"	16"	M	2
R banding	¾"	1½"	21"	M	4
GATE LEGS AND BLOCKS					
S rails	¾"	2¼"	14¼"	M	8
T* legs	1½"	2¼"	31¼"	LM	4
U recessed blocks	½"	2¼"	3⅞"	M	4
VISE MOUNT AND JAWS					
V spacer block	2¼"	5"	9"	LBP	1
W vise jaws	¾"	4"	9"	M	2
X dog-holder	1½"	2¼"	3½"	LM	1

*Initially cut parts marked with an * oversized. Then, trim each to finished size according to the how-to instructions. Note, we measure length with the grain.

Materials Key: BP–birch plywood, M–maple, P–plywood, LM-laminated maple, LBP–laminated birch plywood.

Supplies: #8 x ⅝", #8 x 1", #8 x 1¼", #8 x 1½", and #8 x 2" flathead wood screws, 8–2 x 2" heavy duty hinges (Stanley #814 back flap hinge), 2 sections of 1½" continuous hinge 36" long with #6 x ¾" flathead wood screws, ⅜" dowel stock, 2–½ x 4½" carriage bolts with flat washers and nuts, 2–½ x 3½" lag screws with ½" washers, 2–2½" (3¼" overall height) rigid casters with 8–#8 x ¾" panhead sheet metal screws, 4–⅜" T-nuts, 4–⅜" nuts, 4 adjustable floor glides, clear finish, paint.

continued

Drop-Leaf Mobile Workbench

continued

Build the plywood top

1 From ¾" birch plywood, cut the benchtop center panels (N) and outer leaf panels (O) to the sizes listed in Bill of Materials plus ½" in length and width.

2 Glue and clamp the benchtop panels together in pairs with the edges and ends flush. See the Benchtop drawing for reference. (We used sliding-head type clamps around the perimeter. Then, we drilled and countersunk a few screw holes from the bottom side and added a few screws in the middle to pull the panels tightly together.)

3 Trim the laminated benchtop panels to the finished sizes listed in the Bill of Materials.

4 Rip to width and then miter-cut the banding strips (P, Q, R) to size. Drill countersunk mounting holes, and screw (no glue) them to the laminated benchtops.

The gate-leg assemblies come next

1 Using the dimensions in the Bill of Materials and on the Gate Leg drawing, cut the rails (S) to size.

2 To form the legs (T), cut eight pieces of ¾" maple to 2½" wide by 32" long. Glue four pairs of two pieces each face-to-face, with the edges and ends flush. After the glue dries, scrape it from one edge of each leg. Next, joint or plane that edge flat. Rip the opposite edge of each leg on your tablesaw for a 2¼" finished width. Crosscut both ends of each leg to a 31¼" finished length.

3 Using a dado blade in your tablesaw or radial-arm saw, cut a pair of 2¼" dadoes ¾" deep where shown on the Gate Leg drawing.

4 Checking for square, glue and clamp a pair of rails (S) to each leg (T).

5 On the bottom of each leg, draw diagonals to find center. Drill a 7/16" hole 2" deep at each centerpoint. (We used a brad-point bit to prevent wander.)

6 Tap a ⅜" T-nut into the bottom of each leg, centered over the holes you just drilled. Pry the nuts loose. Mix about an ounce of epoxy, and use an ear swab to coat the portion of each leg bottom and hole that comes into contact with the T-nuts. Tap the T-nuts back into position. After the epoxy cures, thread a ⅜" adjustable floor glide into each T-nut.

7 For locking the leg tops to the bottom surface of the benchtop leaves later, cut the receiver blocks (U) to the size listed in the Bill of Materials and the shape shown on the Receiver Block drawing. Drill the mounting holes. Set the blocks aside for now; we'll add them later.

It's time to add the vise

1 Using the Final Assembly and Spacer Block drawings for reference, laminate three pieces of ¾" plywood to form the spacer block (V). (We used a Record 52-1/2D woodworker's vise with a 9" jaw width. The size of your spacer and its location may vary with different brands of vises. See the Buying Guide at the end of the article for our vise source.)

2 Mark the centerpoints and bore ½" and ⅝" holes through the block where marked on the Spacer Block drawing. Cut from the edge in to the ⅝" holes to form ⅝"-wide slots where shown.

3 Clamp the spacer block (V) to the bottom side of the benchtop center section (N) where shown on the Final Assembly drawing. The outside edge of the block should be flush with the outside edge of the banding strip (Q).

4 Position the vise on the spacer block to determine the block's exact location, and adjust if necessary. With the vise clamped securely to the benchtop, use the existing holes as guides to drill mounting holes through your benchtop. (We clamped scrap stock to the top surface of our benchtop top to prevent chip-out.) Flip the center benchtop section (minus the vise) over, and counterbore the mounting holes.

5 Cut a pair of wood jaws (W) to size, and attach them to the metal vise jaws.

6 Mark the dog-hole centerpoints on the center benchtop section where shown on the Dog Hole Layout drawing accompanying the Benchtop drawing. Drill the holes.

continued

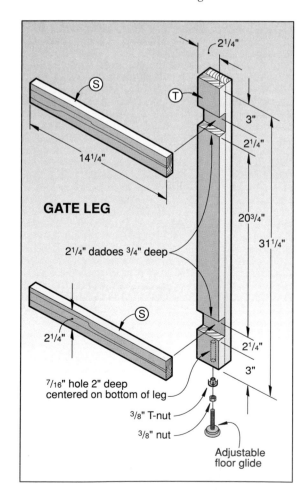

GATE LEG

2¼"

14¼"

2¼"

2¼" dadoes ¾" deep

2¼"

7/16" hole 2" deep
centered on bottom of leg

⅜" T-nut

⅜" nut

Adjustable
floor glide

3"

2¼"

20¾"

31¼"

2¼"

3"

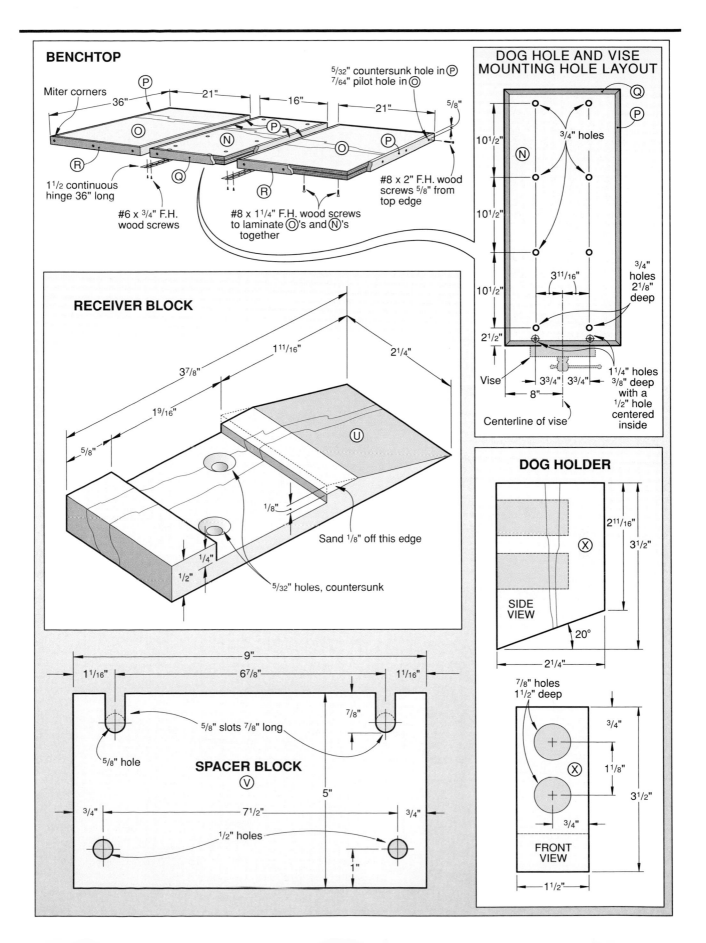

BENCHTOP

Miter corners

36" 21" 16" 21"

5/32" countersunk hole in (P)
7/64" pilot hole in (O)

5/8"

(P) (O) (N) (P) (O) (P) (R)

1 1/2 continuous hinge 36" long

#6 x 3/4" F.H. wood screws

(Q)

#8 x 1 1/4" F.H. wood screws to laminate (O)'s and (N)'s together

(R)

#8 x 2" F.H. wood screws 5/8" from top edge

DOG HOLE AND VISE MOUNTING HOLE LAYOUT

(Q) (P) (N)

3/4" holes

10 1/2"

10 1/2"

10 1/2"

3 11/16"

3/4" holes 2 1/8" deep

2 1/2"

Vise

3 3/4" 3 3/4"

8"

Centerline of vise

1 1/4" holes 3/8" deep with a 1/2" hole centered inside

RECEIVER BLOCK

1 11/16" 2 1/4"

3 7/8"

1 9/16"

5/8"

(U)

1/8"

Sand 1/8" off this edge

1/4"

1/2"

5/32" holes, countersunk

DOG HOLDER

2 11/16" 3 1/2"

(X)

SIDE VIEW

20°

2 1/4"

7/8" holes 1 1/2" deep

3/4"

(X)

1 1/8"

3/4"

3 1/2"

FRONT VIEW

1 1/2"

9"

1 1/16" 6 7/8" 1 1/16"

7/8"

5/8" slots 7/8" long

5/8" hole

SPACER BLOCK
(V)

5"

3/4" 7 1/2" 3/4"

1/2" holes

1"

Drop-Leaf Mobile Workbench

continued

FINAL ASSEMBLY

1 Position a pair of saw horses about 4' apart. Set a pair of 2×4s on the sawhorses and center the benchtop center section upside down on the 2×4s. Position the benchtop leaves next to the center section also upside down. Using two 36"-long sections of 1½" continuous (piano) hinge, drill pilot holes and screw the leaves to the center section.

2 Position the base (also upside down) on the bottom side of the benchtop center section where shown in the photo *below*, centering the base from side-to-side and end-to-end. Place the long cleats (K) on the inside of the side panels (A). See the Final Assembly drawing for reference. Next, clamp the cleats in place to the benchtop. Lift the base off the benchtop, and drive screws through the previously drilled holes in the cleats and into the bottom side of the benchtop.

3 Reposition the base on the benchtop. Drill and countersink mounting holes through the outside face of the base cabinet into the cleats (K). Next, screw the base to the benchtop cleats.

4 Drill the mounting holes and attach the hinges to the ends of the leg rails. Attach the hinge/leg assemblies to the outside face of the side panels. (For proper spacing between the top ends of the legs and the benchtop bottoms, we slipped a receiver block between the two.)

5 Swing the legs 90° from the base sides, and slide the receiver blocks (U) in place under the legs. Use a pencil to trace their location on the benchtop bottom surface. Swing the legs against the base, and screw the receiver blocks in place.

6 Position the handles (upside down) on the K cleats. Screw the handle retainer pieces (L, M) in place.

7 Remove all the hardware and remove the legs from the base and the benchtop assemblies from the base.

8 Cut the dog holder (X) to shape, and drill a pair of holes in it to house your dogs.

9 Sand all assemblies. Apply a clear finish (we used satin polyurethane) to the benchtops and legs. Prime and paint the base and dog holder.

10 Attach the benchtops, legs, and dog holder to the base. Position the handles, and screw the handle retainers (L, M) to the base where shown on the Base Cabinet drawing. Mount the vise and casters.

BUYING GUIDE

• **Vise.** Record 52-1/2D, 9" jaw width, 13" jaw opening, with quick-release mechanism. For the current price contact Seven Corners Hardware Inc., 216 W. 7 Street, St. Paul, MN 55102. Or call 800-328-0457 or 612/224-4859 to order.

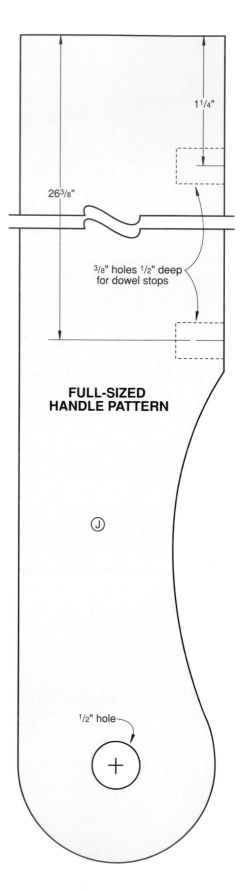

26³/₈"

1¹/₄"

³/₈" holes ¹/₂" deep for dowel stops

FULL-SIZED HANDLE PATTERN

J

¹/₂" hole

Angle-Master Miter Jig

No tablesaw should be without one

To cut accurate miters each and every time without the hassle of making numerous test cuts to constantly verify our miter-gauge setting, we designed this practical jig for the WOOD® magazine shop. And, boy, does it work! The main components included are the one-piece plywood *base* and the two *fences* with adjustable *stops*.

The pivoting twin fences allow us to cut both left and right miters without changing settings. Opposing fences are also necessary to miter-cut molded trim on both ends. Miter jigs with only one fence can perform this operation only on flat stock. In addition, these pivoting fences lock positively without a tendency to creep when tightening the lock knob.

The adjustable stops on the fences provide a range of miter-cutting pieces from 0 to 46" long. The polycarbonate *stop plates* provide a positive stop for mitered ends of stock up to 4" wide. The locks on the aluminum stop bars provide "creep-free" positioning.

For safety, we used a *polycarbonate guard* to provide a dust and chip-free view of the cutting action, and positioned the large *push handle* well above the blade height. We've also added a *safety stop* permanently attached to the side of the tablesaw extension to prevent the jig from moving too far forward and having the saw blade accidently cut through the rear support. Similarly, the sliding table support holds the jig on the table so it doesn't drop off the operator end of the tablesaw when pulled back.

START WITH THE BASE

Note: We built this jig to fit our 10" Delta Unisaw. The width of the base (A) may vary depending on your particular saw. The base should be wide enough to overhang the left-hand saw-table edge by 1½", allowing enough room to attach the safety stops.

1 Using the dimensions on the Parts View drawing on *page 25*, measure carefully, and lay out the outline for the base (A) on a piece of ¾ × 27½ × 40" plywood (we used birch).

2 Mark the centerpoints for all the holes shown on the drawing. Drill the holes to the sizes stated.

3 Cut the base (A) to shape.

continued

Angle-Master Miter Jig

continued

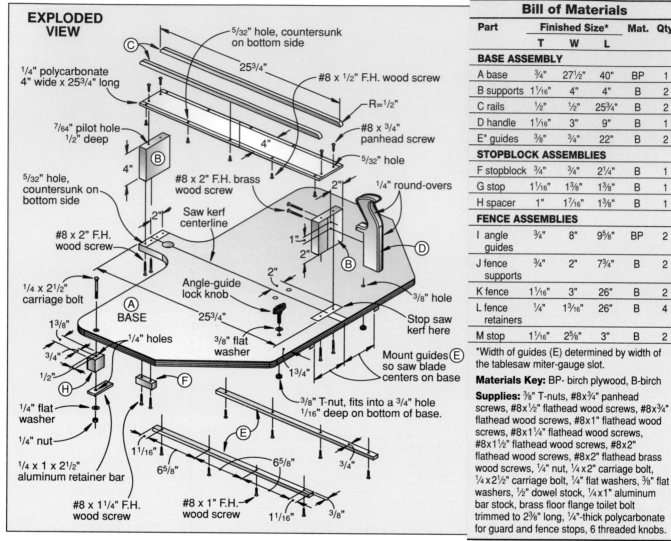

EXPLODED VIEW

¼" polycarbonate 4" wide x 25¾" long

⁵/₃₂" hole, countersunk on bottom side

25¾"

#8 x ½" F.H. wood screw

R=½"

#8 x ¾" panhead screw

⁷/₆₄" pilot hole ½" deep

5/32" hole

4"

4"

⁵/₃₂" hole, countersunk on bottom side

#8 x 2" F.H. brass wood screw

¼" round-overs

2"

2"

Saw kerf centerline

#8 x 2" F.H. wood screw

1"

2"

¼ x 2½" carriage bolt

Angle-guide lock knob

2"

1⅜"

BASE

25¾"

3/8" hole

¼" holes

3/8" flat washer

Stop saw kerf here

¾"

½"

⅜" T-nut, fits into a ¾" hole 1/16" deep on bottom of base.

Mount guides E so saw blade centers on base

1¾"

¼" flat washer

¼" nut

1¹/₁₆"

6⅝"

6⅝"

¾"

¼ x 1 x 2½" aluminum retainer bar

6⅝"

#8 x 1¼" F.H. wood screw

#8 x 1" F.H. wood screw

1¹/₁₆"

3/8"

Bill of Materials

Part	Finished Size*			Mat.	Qty.
	T	**W**	**L**		
BASE ASSEMBLY					
A base	¾"	27½"	40"	BP	1
B supports	1¹/₁₆"	4"	4"	B	2
C rails	½"	½"	25¾"	B	2
D handle	1¹/₁₆"	3"	9"	B	1
E* guides	⅜"	¾"	22"	B	2
STOPBLOCK ASSEMBLIES					
F stopblock	¾"	¾"	2¼"	B	1
G stop	1¹/₁₆"	1⅜"	1⅜"	B	1
H spacer	1"	1⁷/₁₆"	1⅜"	B	1
FENCE ASSEMBLIES					
I angle guides	¾"	8"	9⅝"	BP	2
J fence supports	¾"	2"	7¾"	B	2
K fence	1¹/₁₆"	3"	26"	B	2
L fence retainers	¼"	1³/₁₆"	26"	B	4
M stop	1¹/₁₆"	2⅝"	3"	B	2

*Width of guides (E) determined by width of the tablesaw miter-gauge slot.

Materials Key: BP- birch plywood, B-birch

Supplies: ⅜" T-nuts, #8x¾" panhead screws, #8x½" flathead wood screws, #8x¾" flathead wood screws, #8x1" flathead wood screws, #8x1¼" flathead wood screws, #8x1½" flathead wood screws, #8x2" flathead wood screws, #8x2" flathead brass wood screws, ¼" nut, ¼x2" carriage bolt, ¼x2½" carriage bolt, ¼" flat washers, ⅜" flat washers, ½" dowel stock, ¼x1" aluminum bar stock, brass floor flange toilet bolt trimmed to 2⅜" long, ¼"-thick polycarbonate for guard and fence stops, 6 threaded knobs.

ADD A GUARD FOR SAFETY

1 Cut the front and rear supports (B) to size. Mark and drill a pair of countersunk mounting holes in one support for attaching it to the handle (D) later.

2 Clamp the supports to the base where shown on the Exploded View drawing. Drill mounting holes, and screw (no glue) the supports to the base.

3 Cut the guard top plate to size from ¼" polycarbonate, using the dimensions on the Exploded View drawing. (Although ¼" acrylic would work, we used polycarbonate because of its shatter resistance.)

Mark the locations and drill and countersink mounting holes through the polycarbonate.

4 Cut the two rails (C) to size. Drill mounting holes and screw them to the polycarbonate panel. Then, screw the polycarbonate panel/rails to the supports (B).

5 Transfer the full-sized handle pattern to a piece of 1¹/₁₆" stock, and cut the handle (D) to shape. Rout ¼" round-overs on the handle where shown on the Exploded View drawing.

6 Remove the guard (B, C, and polycarbonate plate) from the base. Clamp the handle to the guard.

Using the previously drilled holes in the rear support as guides, drill ⁷/₆₄" pilot holes into the handle. Screw the handle to the guard. Reattach the guard assembly to the base.

TIME FOR THE MITER GAUGE GUIDES AND STOP

1 Cut a pair of miter-gauge guides (E) to fit snugly into the miter-gauge slots in your tablesaw.

2 Attach the guides to the bottom of the base (A) ensuring that the saw blade is centered on the base and that the base slides back and forth on your tablesaw's slots easily and without wobble. (We rubbed

paraffin wax on the wood guides so the base assembly slides smoothly and effortlessly in the tablesaw miter-gauge slots.)

3 Position the base assembly on your tablesaw, and cut a kerf through the 1¼" hanging hole and *down the center* of the base, stopping just before the blade cuts into the rear support (B).

4 Cut the stop pieces (F, G, H) and ¼×1×2½" aluminum retainer to size.

5 Drill a mounting hole through the stop (G) and your tablesaw table, and bolt part G to your tablesaw table about 1" from the end. See Positioning the Stopblock drawing for reference.

6 Position the jig base (A) on the tablesaw where the saw blade rests at the end of the kerf in the base and next to the rear support (B). Attach the stopblock (F) to the bottom side of the base (A) next to the stop (G) bolted to the saw table. The stopblock/stop combination (F, G) is designed to stop the jig before the blade cuts through the rear support and so the jig doesn't go over the back edge of the tablesaw.

7 Attach the retainer spacer (H) and aluminum retainer to the bottom of the tablesaw base. The spacer/retainer assembly holds the jig horizontal so it doesn't drop off the operator end of the tablesaw when the operator pulls the jig towards himself.

CONSTRUCT THE ANGLE GUIDES

1 Cut two pieces of ¾" birch plywood to 8×10" for the angle guides (I). Tape the two angle-guide blocks face-to-face with double-faced tape.

2 Using the Parts View drawing for reference, lay out and cut the two angle guides (I) to shape.

3 Mark the two ⁷⁄₁₆" hole centerpoints on the taped-together pieces. Use a compass to

continued

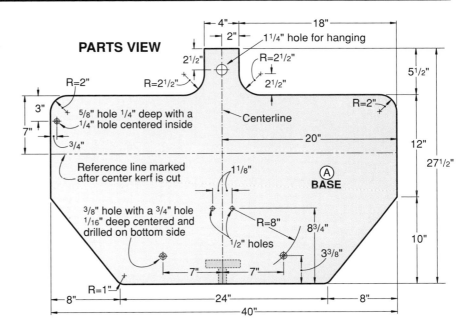

PARTS VIEW

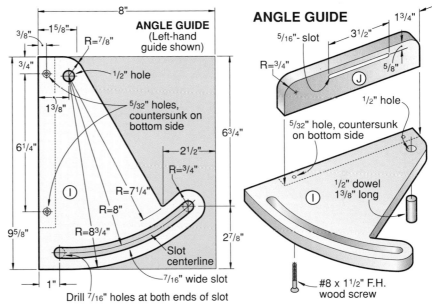

ANGLE GUIDE (Left-hand guide shown)

ANGLE GUIDE

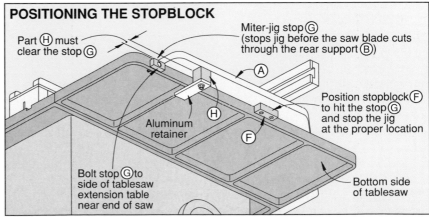

POSITIONING THE STOPBLOCK

Angle-Master Miter Jig

continued

mark the outline of the $7/16$"-wide slot between the $7/16$" holes. Drill the $7/16$" holes at each end of the slot. Using a scrollsaw or jigsaw cut along the marked lines between the holes to form the slot. Sand the slot edges.

4 Use a wood wedge to pry the pieces apart. If the tape bond resists, use lacquer thinner or acetone to weaken the tape.

5 Cut the fence supports (J) to shape, and form a $5/16$"-wide slot in each. Screw a fence support to each angle guide. As noted on the drawing, the left-hand guide is shown; the right-hand guide is a mirror image.

6 Crosscut a pair of $1/2$" dowels to $1\frac{3}{8}$" long. Glue one in each angle guide (I) where shown on the Angle Guide drawing.

LET'S ADD THE FENCE ASSEMBLIES AND STOPS

1 Cut the two fence pieces (K) and four retainers (L) to size from solid stock.

2 Mark the centerpoint on each fence piece (K), and drill a $1/2$" hole $1/2$" deep with a $1/4$" hole centered inside where shown on the Fence drawing for the $1/4 \times 2$" carriage bolt. The carriage bolt that goes through this hole will be used to connect the fence (K) to the fence support (J).

3 To drill the hole and recess for the stop lock-down mechanism (we're using a brass floor flange toilet bolt trimmed to $2\frac{3}{8}$" long for this), follow the procedure in the Hole/Recess detail *at left*.

4 Cut or rout a 1" groove $1/4$" deep centered along the length of each fence piece (K) where shown on the Fence drawing *below*.

5 Glue and clamp the retainers (L) to the

HOLE/RECESS DETAIL

$1/4$" hole

Recess

Side Section view of (K)

$1\frac{1}{2}$"

$3/8$"
$5/16$"
$3/8$"

Bottom view of (K)

STEP 1
Drill two $1/2$" holes $1\frac{1}{2}$" deep on centerpoints

STEP 2
Chisel sides of recess

STEP 3
Drill a $1/4$" hole from top

FENCE
(Left-hand fence)

Threaded knob

$1/4$" flat washer

$1/2$" hole $1/2$" deep with a $1/4$" hole centered inside, drilled before cutting groove in (J)

$1/4$" flat washer

Threaded knob

1" groove $1/4$" deep

1"

$5/8$"

1"

$1\frac{3}{16}$"

$5/8$"

(L)

$1\frac{3}{16}$"

Brass toilet bolt

(K)

$7/8$"

$5\frac{1}{8}$"

26"

Sand or file end of toilet bolt if necessary to fit into recess in bottom of (K)

$1/4 \times 2$" carriage bolt

Note: Insert bolt into hole in (K) before gluing (L) to (K)

HANDLE PATTERN

(D)

fences (K) keeping the inside edges of the guides ⅝" apart the length of the fence. See the Fence Assembly Section View for reference.

6 Using the Fence Stop drawing for reference, cut the stops (M) to size. (For safety when machining, we started with a piece 1¹⁄₁₆×2⅝×10" and cut a pair of 1" rabbets ¼" deep along one face. Then, we crosscut two 3"-long sections from it.) Check the fit of the the stops (M) between the retainers (L). The stop should slide back and forth easily without an excess of free play.

7 Cut two ¼×3×4" pieces of polycarbonate to act as stop plates. Drill a pair of countersunk mounting holes in each, and screw the stop plates to the bottom of the stops (M) where shown on the Fence Stop drawing.

8 Crosscut two pieces of ¼×1" aluminum bar stock to 25" long. Drill a countersunk pair of mounting holes in each bar where shown on the Fence Stop drawing. Screw the stops (M) to the aluminum stock.

9 Trim a brass floor flange toilet bolt to 2⅜" long, and insert it into the recessed hole in the bottom side of fence (K) where shown on the Fence drawing. You may need to sand or file the head of bolt slightly to fit into the recess. Slide a washer onto the top of the toilet bolt, and add a threaded knob. Turn the knob clockwise to pull the brass toilet bolt

up against the bottom of the aluminum bar, locking it in place.

MARKING THE ANGLE REFERENCE MARKS

1 As shown in Photo A *below*, position and clamp a straightedge perfectly square to your saw blade on the jig base. (As shown in the photo, we used an adjustable triangle to accurately locate the straightedge. See the Buying Guide for our source of this handy device.) Mark a reference line across the jig where shown on the Parts View drawing. Leave the straightedge clamped to the base.

2 Position the pivoting fences (minus the stops) on the base.

3 Align both fences parallel to the marked reference line. Mark a

second reference line along the *front edge* of the fences. Use this line to position either fence for crosscutting at 90°.

continued

FENCE STOP

25"
1"
1½"
¾"
½"
¼" aluminum bar 1" wide x 25" long
3"
2⅝"
⅝"
1"
#8 x ¾" F.H. wood screw
¼"
5/32" hole, countersunk on back side
7/16"
4"
¾"
5/32" holes, countersunk on bottom side. Mating holes in bottom of (M) are 7/64" pilot holes ½" deep
¼ x 3 x 4" polycarbonate
#8 x ¾" F.H. wood screw

END VIEW DETAIL
Sand a slight chamfer on edges of aluminum
3/16"
Aluminum bar

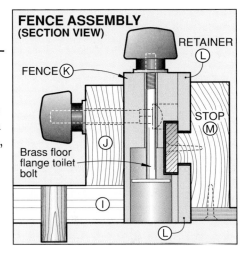

FENCE ASSEMBLY (SECTION VIEW)
RETAINER L
FENCE K
STOP M
Brass floor flange toilet bolt
J
I
L

A

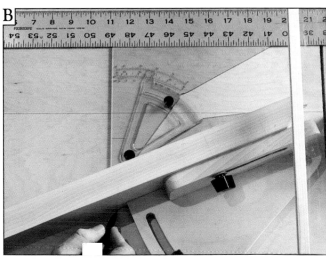

B

Angle-Master Miter Jig

continued

4 As shown in Photo B *on page 27*, use the adjustable triangle to accurately position the fence. Then, mark reference lines onto the base, using the *back edge* of the fence as a straightedge. (We marked reference lines at 15°, 22.5°, 30°, and 45°.) Repeat for both sides of the jig.

HOW TO USE YOUR NEW JIG

Review the four drawings *below* to determine our method for making miter-cuts and crosscuts.

BUYING GUIDE

• **Miter jig hardware kit.** For current prices of screws, toilet bolts, aluminum bar stock for fences and safety stop, polycarbonate for guard and stops, nuts, bolts, and washers, Kit no. WM–MJ, contact Miller Hardware, 1300 M.L. King Pkwy., Des Moines, IA 50314 or call 515/283-1724 to order.

• **Adjustable triangle.** For current price of an 8" model, stock no. 8211-8, contact The Art Store, 600 M.L. King Pkwy., Des Moines, IA 50312 or call 800/652-2225 or 515/244-7000 to order.

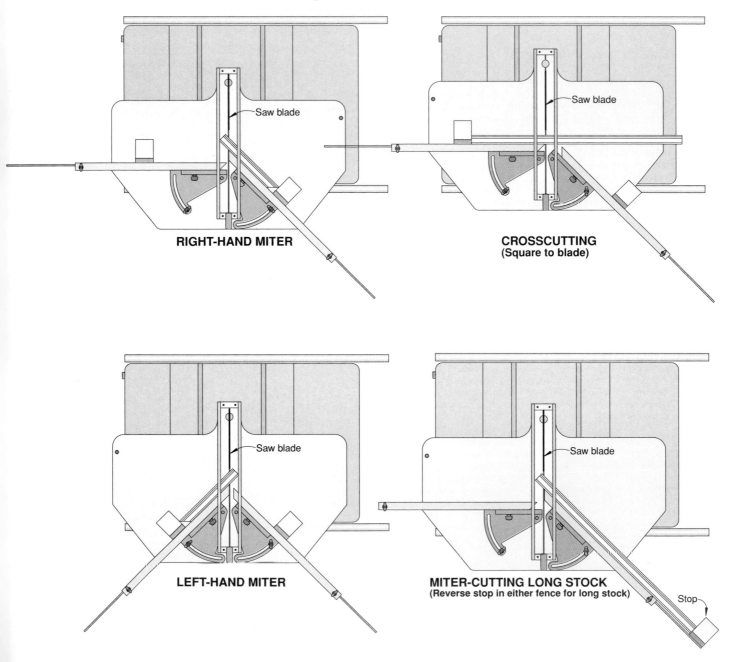

RIGHT-HAND MITER

Saw blade

CROSSCUTTING
(Square to blade)

Saw blade

LEFT-HAND MITER

Saw blade

MITER-CUTTING LONG STOCK
(Reverse stop in either fence for long stock)

Saw blade

Stop

Finishing Center

Now you see it, now you don't

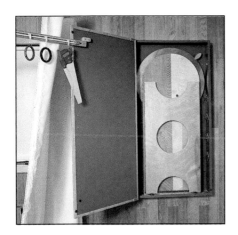

When we designed the WOOD® magazine IDEA SHOP, floor space was at a premium. That's why we decided to go with the wall-hung cabinet shown here. It features a fold-down worktable that rotates on a lazy Susan bearing, allowing you to apply an even coat of finish on all sides of your project without repositioning it. And, a support arm attached to the cabinet door lets you do the same with small projects or parts. Then, when you're through for the day, just lift up on the worktable, stash it back in the cabinet, and close the door.

START WITH THE PLYWOOD CABINET ASSEMBLY

1 From ¾" maple, cut the cabinet sides (A) and top and bottom (B) to the sizes listed in the Bill of Materials on the *page 30*.

2 Cut a ¾" rabbet ⅜" deep across the ends of the cabinet sides (A).

3 Switch to a ½"-wide dado blade, and cut a ½" rabbet ⅜" deep along the back inside edge of all four cabinet pieces (A, B).

4 Dry-clamp the cabinet pieces, measure the opening, and cut the back (C) to size from ½" plywood. Remove the clamps.

5 Glue and clamp the basic cabinet (A, B) together, checking for

square. While the glue is still wet, attach the back (C) with 4d finish nails and glue. The plywood back helps hold the unit square until the glue dries. Remove any glue from the inside of the cabinet with a damp cloth. For added strength, drive a pair of 6d nails at each corner where shown on the Exploded View drawing. Set the nails and fill the holes with putty.

6 To hold the drop-down table assembly in the cabinet later, cut

the spacer (D) and knob (E) to shape. See the Parts View drawing for the full-sized knob pattern. Drill a 5/32" hole centered in the spacer. Drill and countersink the same sized hole in the knob.

7 Mark the hole centerpoints, and drill a pair of holes for the round magnetic catches and wood screws in the cabinet sides (A) where shown on the Exploded View drawing. You'll add the catches, knob, spacer, and strike plates later.

continued

Finishing Center

continued

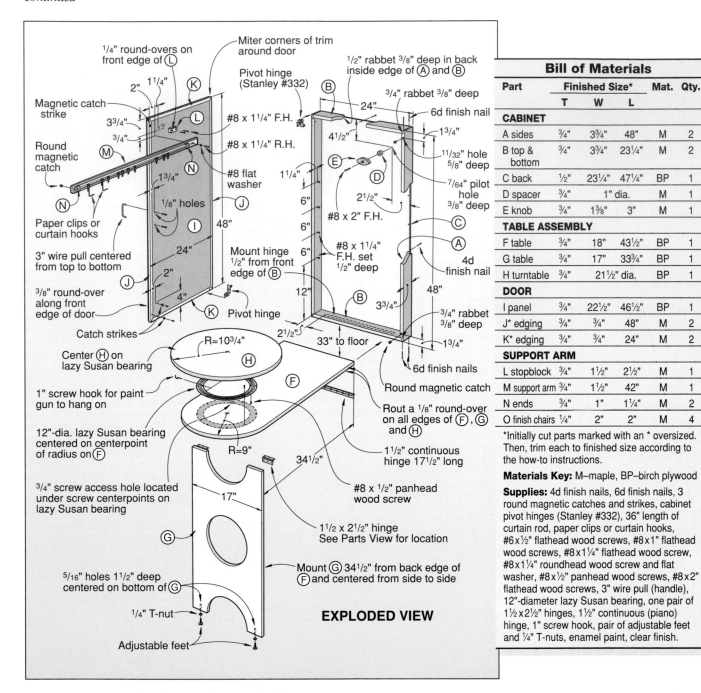

Bill of Materials

Part	Finished Size*			Mat.	Qty.
	T	W	L		
CABINET					
A sides	3/4"	3³/4"	48"	M	2
B top & bottom	3/4"	3³/4"	23¹/4"	M	2
C back	1/2"	23¹/4"	47¹/4"	BP	1
D spacer	3/4"	1" dia.		M	1
E knob	3/4"	1³/8"	3"	M	1
TABLE ASSEMBLY					
F table	3/4"	18"	43¹/2"	BP	1
G table	3/4"	17"	33³/4"	BP	1
H turntable	3/4"	21¹/2" dia.		BP	1
DOOR					
I panel	3/4"	22¹/2"	46¹/2"	BP	1
J* edging	3/4"	3/4"	48"	M	2
K* edging	3/4"	3/4"	24"	M	2
SUPPORT ARM					
L stopblock	3/4"	1¹/2"	2¹/2"	M	1
M support arm	3/4"	1¹/2"	42"	M	1
N ends	3/4"	1"	1¹/4"	M	2
O finish chairs	1/4"	2"	2"	M	4

*Initially cut parts marked with an * oversized. Then, trim each to finished size according to the how-to instructions.

Materials Key: M–maple, BP–birch plywood

Supplies: 4d finish nails, 6d finish nails, 3 round magnetic catches and strikes, cabinet pivot hinges (Stanley #332), 36" length of curtain rod, paper clips or curtain hooks, #6x½" flathead wood screws, #8x1" flathead wood screws, #8x1¼" flathead wood screw, #8x1¼" roundhead wood screw and flat washer, #8x½" panhead wood screws, #8x2" flathead wood screws, 3" wire pull (handle), 12"-diameter lazy Susan bearing, one pair of 1½x2½" hinges, 1½" continuous (piano) hinge, 1" screw hook, pair of adjustable feet and ¼" T-nuts, enamel paint, clear finish.

EXPLODED VIEW

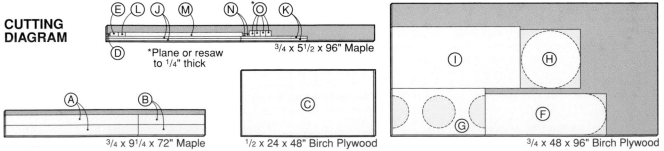

CUTTING DIAGRAM

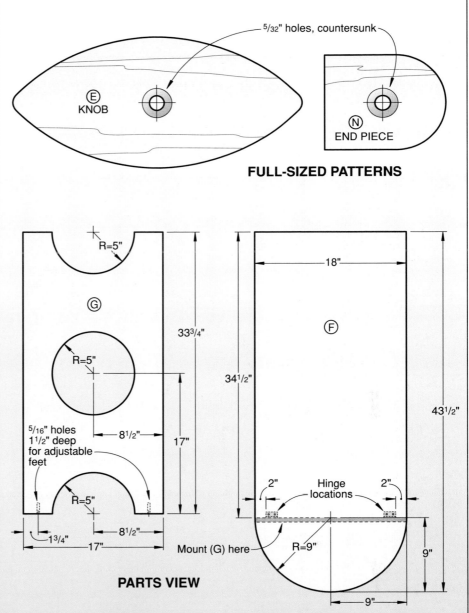

FULL-SIZED PATTERNS

⁵/₃₂" holes, countersunk

Ⓔ KNOB

Ⓝ END PIECE

PARTS VIEW

R=5"

Ⓖ

R=5"

33³/₄"

8¹/₂"

17"

⁵/₁₆" holes 1¹/₂" deep for adjustable feet

R=5"

1³/₄"

8¹/₂"

17"

18"

Ⓕ

34¹/₂"

43¹/₂"

2"

Hinge locations

2"

Mount (G) here

R=9"

9"

9"

NOW, LET'S ADD THE DROP-DOWN TABLE

1 Cut the table (F), table support (G), and turntable (H) to size and shape. See the Parts View drawing for reference when marking the layout for parts F and G.

2 Using the radius centerpoint you laid out for the end of the table (F), center a 12"-diameter lazy-Susan bearing on top of the table (F). Use two screws to temporarily hold the bearing to the table top. Next, mark the centerpoint, and drill a screw-access hole through the table. The access hole in the table allows you to work from the bottom of the table later to screw the lazy Susan bearing to the bottom side of the turntable (H).

3 Drill mounting holes in the bottom edge of the table support (G) for the adjustable feet (see the Exploded View drawing).

NOW A DOOR TO KEEP THE DUST OUT

1 Cut the door panel (I) to size. Cut the door edging (J, K) to size plus 2" in length. Then, miter-cut the stiles and rails, and glue and nail them to the door panel. Drill the wire pull (handle) mounting holes through the door.

2 Drill the mounting holes and add a pair of pivot hinges to the cabinet and door to check the fit. See the Mounting The Pivot Hinges drawing and the hinge manufacturer's instructions for reference. Remove the hinges for now.

ADD THE SUPPORT ARM AND THE FINISH CHAIRS

1 Cut the stopblock (L), support arm (M), and end pieces (N) to size and shape. See the Parts View for help with the end pieces (N).

2 Sand ⅛" round-overs along the front edges of all four pieces.

3 Trim a section of curtain rod to 36" long, and secure it to the support arm (M) where shown on the Hanger Assembly drawing.

4 Screw the support/curtain rod to the door. Do not overtighten the screw. The support arm must be free to swing from a vertical to a horizontal position.

5 Now, attach the stopblock (L) to the door. The stopblock should be positioned so that when the support arm is horizontal, the end of the support arm can slide under the stopblock and be held up to the inside face of the door where shown on the Exploded View drawing.

6 Build four finish chairs (O) as shown on the Finish Chair drawing. As shown in the opening photograph, the chairs hold projects above the turntable for ease in applying finishes. Drill mounting holes in the right or left-hand side piece (A), and drive four screws on which to hang the chairs.

FINISH IT LIKE A PRO, AND ASSEMBLE THE PIECES

1 Remove all the hardware from the cabinet, table assembly, and door. Finish-sand the project.

2 Mask the surrounding areas, and apply a clear finish to the door

continued

Finishing Center

continued

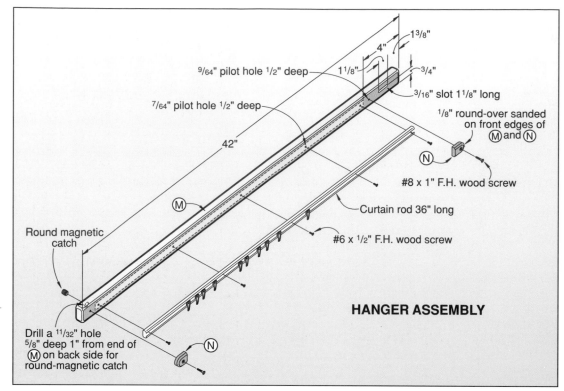

9/64" pilot hole 1/2" deep

7/64" pilot hole 1/2" deep

42"

13/8"

4"

11/8"

3/4"

3/16" slot 11/8" long

1/8" round-over sanded on front edges of Ⓜ and Ⓝ

#8 x 1" F.H. wood screw

Curtain rod 36" long

#6 x 1/2" F.H. wood screw

Round magnetic catch

Ⓜ

Ⓝ

Ⓝ

HANGER ASSEMBLY

Drill a 11/32" hole 5/8" deep 1" from end of Ⓜ on back side for round-magnetic catch

Ⓝ

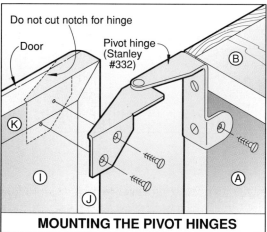

Do not cut notch for hinge

Door

Pivot hinge (Stanley #332)

Ⓑ

Ⓚ

Ⓘ

Ⓙ

Ⓐ

MOUNTING THE PIVOT HINGES

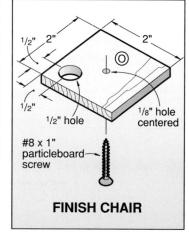

1/2"

2"

2"

Ⓞ

1/2"

1/2" hole

1/8" hole centered

#8 x 1" particleboard screw

FINISH CHAIR

the table support (G) to the bottom of the table (F) where shown on the Exploded View drawing.

8 Trim a 17½"-long section of 1½" continuous (piano) hinge. Drill the pilot holes and fasten the hinge to the back bottom edge of the table.

9 To mount the cabinet to the wall, locate the stud(s), and with the aid of a helper, position the cabinet on the wall 33" from the floor. (We cut a 2×4 to length, and used it as a temporary leg to hold the cabinet at a certain height against the wall. Level the cabinet, and drill and countersink mounting holes through the cabinet back and into the wall stud(s). Screw the cabinet to the wall.

10 Use the pivot hinges to reattach the door to the cabinet.

11 With the table support at a right angle to the table and the adjustable feet of the table support on the shop floor, screw the continuous hinge to the bottom cabinet frame member (B). If the table (F) doesn't sit level, adjust the feet first; if this doesn't provide enough adjustment, raise or lower the cabinet as needed.

12 Raise the table assembly (F, G, H) into the cabinet. Now, drill a pilot hole and attach the spacer and knob (D, E) in a position so when rotated, the knob will hold the assembly in place.

edging, table assembly, and support arm. Later, mask the areas that received a clear finish, and paint the cabinet and door.

3 Insert a pair of magnetic catches into the holes in the right-hand cabinet side (A). Drill and fit a catch into the hole in the back side of the support arm.

4 Secure the curtain rod to the support arm. For hanging projects later, add paper clips to the rod. Secure the stopblocks (N) to the end

of the rods to keep the curtain hooks from sliding out.

5 Mark its mating location and add the strike plate to the inside face of the door. This keeps the support arm from swinging back and forth every time the cabinet door is opened and closed.

6 Fasten the curtain rod assembly to the door where shown on the Exploded View drawing.

7 Drill the pilot holes and use a pair of 1½ x 2½" hinges to mount

Furniture With Finesse

Nothing satisfies more than seeing a fine piece of furniture and knowing it came from your own shop. We select our furniture designs not only for their utility, but also on how good they look. Build one of these projects and we know you'll get the compliments you deserve.

Comfy Country Chair

Affordable, easy-to-build seating for your deck or patio

This simply styled furniture piece proves that you don't have to spend a fortune to spruce up your favorite outdoor area. If you're interested in expanding this country piece into a set of furniture, take a look at the Picnic-Perfect Table and Benches on *page 38*, and the photograph and Buying Guide for the Country Comfy Trio on *page 41*.

Note: *For our chair, we hand-picked fir 2× stock. Pine, spruce, or redwood also will work well. If you have trouble locating straight and uncupped stock, edge-join narrower pieces to width. For joints that will stand up to the extremes of Mother Nature, use Titebond II water-resistant glue, slow-set epoxy, or resorcinol glue.*

BUILD THE CHAIR ENDS FIRST

1 Cut the ends (A) to 20¾" long from 2×12 or edge-joined stock.

2 Transfer the full-sized heart half-pattern onto a piece of heavy paper or poster board. Cut the full-sized template to shape.

3 Position the template, and trace the heart outline on all four end pieces (A) where located on the End View drawing. Cut the marked outlines to shape on the bandsaw or with a jigsaw, and then drum-sand the cut edges to remove the saw marks.

4 Clamp each matching pair of 2×12s together, heart edge to edge, with the top and bottom edges flush. Now, using the dimensions on the Exploded View drawing, mark the two dowel-hole locations on one face. Remove the clamps. Using a square, transfer the lines to the inside edge of each end piece.

5 Check that you're square to the edge, and bore ¾" holes 1½" deep centered from edge to edge where marked. (We used an electric drill and a spade bit.)

6 Set a stop, and crosscut four pieces of ¾" oak dowel stock to 3⅜" long. For ease of insertion, sand a chamfer on each end of each dowel. (We formed our chamfers on a belt sander.)

7 Cut four ⅜"-thick spacers. Glue, dowel, and clamp both chair ends together, placing the ⅜" spacers between the end pieces (A) for a consistent ⅜" gap. Save the spacers; you'll use them when joining the seat and back pieces later.

8 Using trammel points, swing an arc to mark the 15⅜" radius on the bottom end of each chair assembly where dimensioned on the End View drawing. Cut the arcs to shape.

9 Sand a slight round-over on all edges of each chair end assembly.

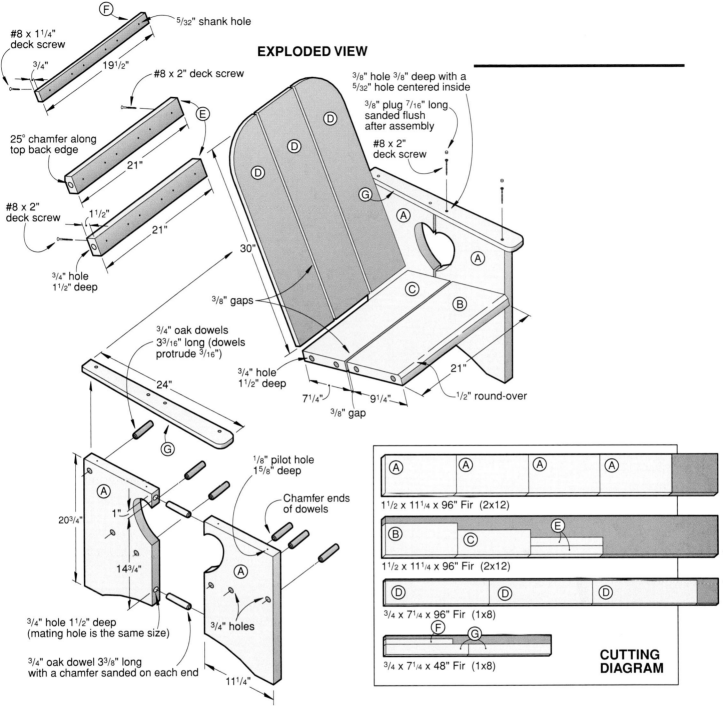

EXPLODED VIEW

#8 x 1¼" deck screw

5/32" shank hole

F

3/4"

19½"

#8 x 2" deck screw

E

25° chamfer along top back edge

21"

#8 x 2" deck screw

1½"

21"

3/4" hole 1½" deep

3/8" hole 3/8" deep with a 5/32" hole centered inside

3/8" plug 7/16" long sanded flush after assembly

#8 x 2" deck screw

D
D
D
D
D

G

A

A

30"

C

B

3/8" gaps

3/4" oak dowels 3³/16" long (dowels protrude 3/16")

24"

G

3/4" hole 1½" deep

7¼"

9¼"

3/8" gap

21"

½" round-over

1/8" pilot hole 1⁵/8" deep

Chamfer ends of dowels

A

1"

20¾"

14¾"

3/4" hole 1½" deep (mating hole is the same size)

3/4" holes

A

3/4" oak dowel 3³/8" long with a chamfer sanded on each end

11¼"

CUTTING DIAGRAM

A | A | A | A
1½ x 11¼ x 96" Fir (2x12)

B | C | E
1½ x 11¼ x 96" Fir (2x12)

D | D | D
3/4 x 7¼ x 96" Fir (1x8)

F | G
3/4 x 7¼ x 48" Fir (1x8)

NOW, FOR THE SEAT

1 Cut the two seat pieces (B, C) to length. (We ripped the two pieces to width from 2×12 stock.)

2 Rout a ½" round-over along the top front edge of the seat front piece (B).

BUILD THE BACKREST NEXT

1 Cut the backrest splats (D) and backrest cleats (E, F) to size.

2 Mark a 6" radius on two of the backrest splats where shown on the Chair Back drawing. Cut the corners to shape and sand smooth to remove the saw marks.

3 To keep the back edge of the middle cleat (E) flush with the back edge of the chair ends (A), bevel-rip a 25° chamfer along the top edge of the cleat where shown on the End View drawing.

continued

Bill of Materials

Part	Finished Size*			Mat.	Qty.
	T	W	L		
A ends	1½"	11¼"	20¾"	C	4
B seat front	1½"	9¼"	21"	C	1
C seat rear	1½"	7¼"	21"	C	1
D splats	3/4"	6¾"	30"	C	3
E cleats	1½"	2½"	21"	C	2
F cleat	3/4"	1½"	19½"	C	1
G armrests	3/4"	3"	24"	C	2

Material Key: C–choice (fir, pine, spruce, or redwood)

Supplies: 2–36" lengths of 3/4" oak dowel stock, #8x1¼" deck screws, #8x2" deck screws, primer, exterior-grade stain or paint.

Comfy Country Chair

continued

4 Using the dimensions on the Exploded View and Chair Back drawings, clamp the cleats (E, F) against the splats (D), using $\frac{3}{8}$" spacers to create gaps between the splats. Check for square.

MARK AND DRILL THE DOWEL HOLES

1 Using the dimensions on the End View drawing, mark the seat centerline first and then the centerline for the backrest cleats on the *outside face* of each seat end assembly. Locate and mark the six dowel-hole centerpoints on the marked lines on each chair end.

2 Bore $\frac{3}{4}$" holes through the chair ends at the marked centerpoints, backing the stock with scrap to prevent chip-out.

ASSEMBLE THE PIECES

1 From $\frac{3}{4}$"-diameter oak dowel stock, set a stop, and cut 12 dowels to $3\frac{3}{16}$" long. Sand a $\frac{3}{16}$" chamfer on both ends of each dowel.

2 Cut two 1×2 scraps to 26" long and two to 14" long. Clamp one of each length to the inside face of each chair end where shown on the Support Locations drawing. The strips help center the ends of the seat and backrest assembly pieces over the $\frac{3}{4}$" holes. (To test the locations, we positioned a piece of 2× stock on each support to check that the holes in the end pieces would center on the ends of the seat pieces (B, C) and cleats (E) before drilling.)

3 With a helper, position the seat pieces where located on the End View drawing. Slip the $\frac{3}{8}$" spacers between the pieces for a consistent gap. Clamp the seat pieces firmly between the chair ends.

4 Chuck a $\frac{3}{4}$" spade bit into a portable electric drill. Using the previously bored holes in the end sections as guides, bore a pair of $1\frac{1}{2}$"-deep holes squarely into each

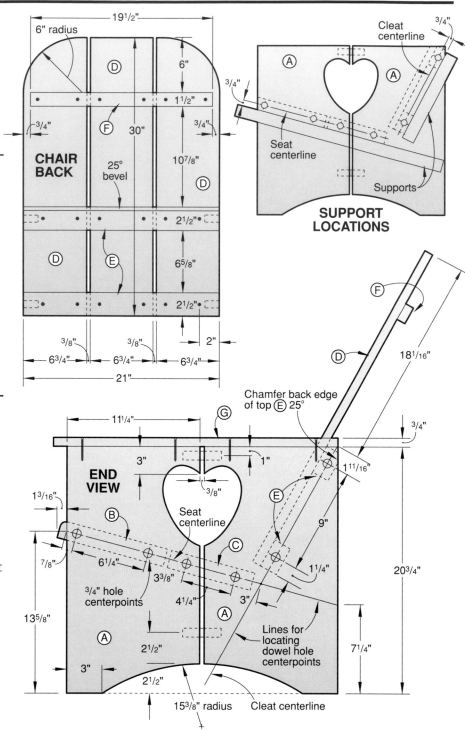

seat piece end. As soon as you've bored the first hole, insert one of the $3\frac{3}{16}$"-long dowels into the hole to help steady the seat piece for boring the next hole. Do not insert the dowel more than $\frac{1}{2}$" into the seat

piece; you may have trouble removing it if you insert it all the way.

5 Repeat the procedure to position and drill the $\frac{3}{4}$" holes in both ends of the backrest cleats (E).

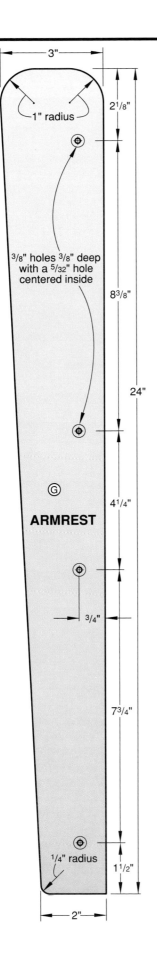

3"

2¹⁄₈"

1" radius

³⁄₈" holes ³⁄₈" deep
with a ⁵⁄₃₂" hole
centered inside

8³⁄₈"

24"

Ⓖ

ARMREST

4¹⁄₄"

3⁄4"

7³⁄₄"

¹⁄₄" radius

1¹⁄₂"

2"

6 Remove one of the 3³⁄₁₆"-long dowels. With a small brush, coat the inside of the hole with glue. To prevent marring the chamfered dowel end, use a rubber-tipped mallet to slowly drive the dowel into the hole. Drive the dowel until just the chamfered end protrudes. Be careful not to drive the dowels too far—they're almost impossible to back out. Immediately wipe off any excess glue. Repeat for each dowel. Let the glue dry and then remove the clamps.

NOW, LET'S ADD THE ARMRESTS

1 From ¾" stock, cut two pieces to 3×24" for the armrests (G).

2 Using the dimensions on the armrest drawing, mark the profile on one piece and cut it to shape. Use the first piece as a template to mark the shape onto the second armrest, and then cut it to shape.

3 Mark the hole centerpoints. Drill and counterbore the holes.

4 Screw the armrests to the tops of the end assemblies (A).

5 Plane or resaw a piece of stock to ⁷⁄₁₆" thick, and use a plug cutter to cut ⅜"-diameter plugs. Plug the holes, and sand the plugs flush with the top of each armrest.

SAND, PAINT, AND SIT A SPELL

1 Sand the entire chair, sanding a slight round-over on all edges.

2 If you decide to paint your chair, an oil-based enamel or water-based latex will provide adequate protection. Regardless of your paint selection, be sure to apply a prime coat that's compatible with the top coat. Also, be sure to apply several coats to the porous end grain.

For a more natural look, finish the chair with an exterior house stain, and then apply several coats of spar varnish.

FULL-SIZED HEART HALF-PATTERN

Centerline

Picnic-Perfect Table and Benches

A simply irresistible country combo

To help you get ready for those summertime picnics, here's a winning combination that won't take but a weekend or two to build.

Note: *For our picnic table and benches, we hand-picked fir 2× stock. Pine, spruce, or redwood also will work well. For joints that will stand up to the* extremes of Mother Nature, use Titebond II water-resistant glue, slow-set epoxy, or resorcinol glue.

BEGIN WITH THE TABLE END PANELS

1 From 2×8 stock, cut the end uprights (A) to length.

2 Transfer the larger full-sized heart half-pattern onto a piece of heavy paper or poster board. Cut the full-sized template to shape.

3 Position the template, and trace the heart outline on all four uprights (A) 10⅞" from the bottom edge of each where located on the Table End Section View drawing. Cutting just inside the marked outlines, cut the half patterns to shape on a bandsaw or jigsaw. Then, drum-sand to the line to remove saw marks.

Bill of Materials

Part	Finished Size*			Mat.	Qty.
	T	W	L		
TABLE					
A uprights	1½"	7¼"	23½"	2x8	4
B feet	1½"	3½"	25"	2x4	2
C top rails	1½"	3½"	25"	2x4	3
D stretchers	1½"	3½"	51"	2x4	2
E tabletop	1½"	5½"	72"	2x6	5

Materials Key: choice of fir, pine, spruce, or redwood

Supplies: ¾" dowel stock, 2½" deck screws, 3" deck screws, 3¾" deck screws

4 Cut the feet (B) and top rails (C) to the lengths listed in the Bill of Materials from 2×4 stock.

5 Mark the cutlines and cut the ends and bottom of the feet to shape. Chamfer the ends of the tops rails where shown.

6 Clamp each matching pair of uprights (A) together, heart edge to edge, with the top and bottom edges flush and a couple of ½" spacers between the parts. Using the dimensions on the Table End Section View drawing, mark the dowel-hole centerlines on the bottom ends of the uprights. Using a square, transfer the dowel-hole centerlines to the mating top edges of the feet. Drill the mating holes in the top edges of the feet (we used a spade bit).

7 Crosscut eight 3" lengths of ¾" dowel stock. For ease of insertion, sand a chamfer on both ends of each dowel.

8 Glue, dowel, and clamp each pair of uprights (A) to the feet (B).

9 Clamp the top rails to the top end of the uprights. Drill the mounting holes, and screw the rails to the uprights. Set the middle top rail (C) aside for now; you'll add it later.

continued

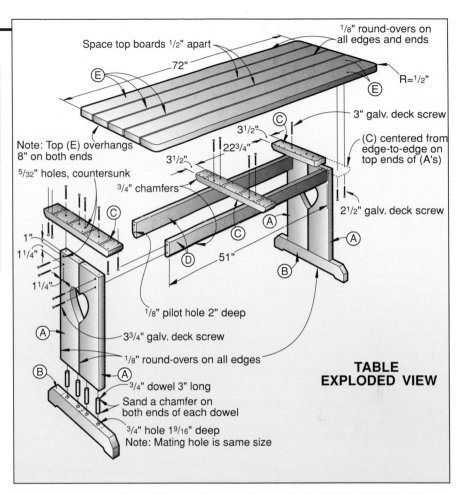

Space top boards ½" apart
72"
⅛" round-overs on all edges and ends
R=½"
3" galv. deck screw
(C) centered from edge-to-edge on top ends of (A's)
3½"
22¾"
3½"
¾" chamfers
2½" galv. deck screw
Note: Top (E) overhangs 8" on both ends
5/32" holes, countersunk
1"
1¼"
1¼"
⅛" pilot hole 2" deep
3¾" galv. deck screw
⅛" round-overs on all edges
¾" dowel 3" long
Sand a chamfer on both ends of each dowel
¾" hole 1 9/16" deep
Note: Mating hole is same size
51"

TABLE EXPLODED VIEW

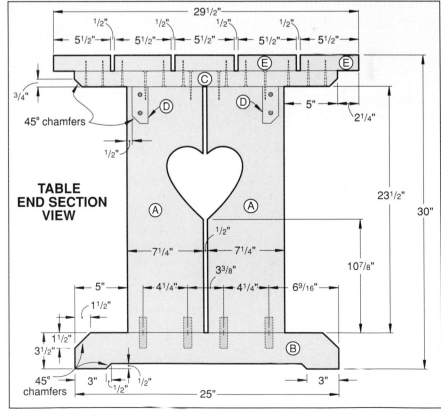

29½"
½" ½" ½"
5½" 5½" 5½" 5½" 5½"
¾"
45° chamfers
½"
5"
2¼"
23½"
30"
TABLE END SECTION VIEW
7¼" 7¼"
½"
3⅜"
10⅞"
5" 4¼" 4¼" 6 9/16"
1½"
1½"
3½"
45° chamfers
3" ½" ½"
25"
3"

Picnic-Perfect Table and Benches

continued

ADD THE STRETCHERS AND TABLETOP PIECES

1 Cut the stretchers (D) to length from 2×4 stock and the tabletop pieces (E) from 2×6 stock.

2 Bevel-rip a 45° chamfer along one edge of each stretcher where shown on the drawing on *page 39.*

3 Rout or sand ⅛" round-overs along all edges and ends of the tabletop pieces.

4 Clamp the stretchers between the end panels where shown on the Table End Section View and Table Exploded View drawings. Drill mounting holes and screw the stretchers in place to complete the base assembly.

5 Position the tabletop pieces *good face down* on sawhorses or on your workbench. Using ½" spacers between the 2×6s, clamp the tabletop pieces together with the ends being flush. Position the base assembly, also upside down, centered from side-to-side and end-to-end on the tabletop pieces. Now, clamp the top rails (C) to the tabletop pieces. Drill mounting holes and screw the two assemblies together. Clamp the remaining top rail (C) centered on the bottom of the tabletop pieces. Drill the pilot holes and screw it in place. Position the table upright on the floor.

6 Mark and then cut a ½" radius on each of the four corners of the tabletop. Sand each corner smooth to remove the saw marks. Rout ⅛" round-overs on the four radiused corners.

THE BUILT-TO-LAST BENCHES COME NEXT

To make the benches, use the same construction process used to build the picnic table. See the Benches Bill of Materials for sizes. Note that the narrower feet (F) on the benches allowed us to go with screws instead of dowels to secure the feet to the

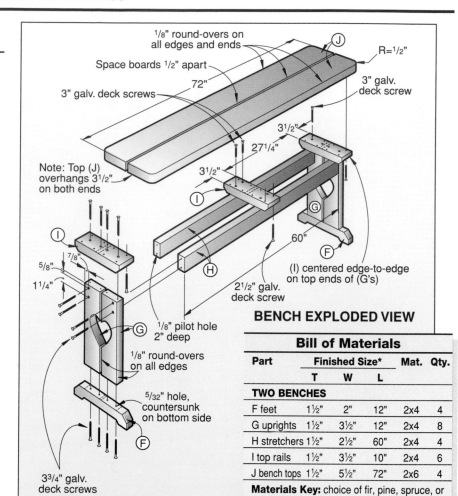

1/8" round-overs on all edges and ends

Space boards ½" apart

3" galv. deck screws

72"

R=1/2"

3" galv. deck screw

3½"

27¼"

3½"

Note: Top (J) overhangs 3½" on both ends

7/8"

5/8"

1¼"

1/8" pilot hole 2" deep

1/8" round-overs on all edges

5/32" hole, countersunk on bottom side

3¾" galv. deck screws

2½" galv. deck screws

60"

(I) centered edge-to-edge on top ends of (G's)

BENCH EXPLODED VIEW

Bill of Materials					
Part	**Finished Size***			**Mat.**	**Qty.**
	T	**W**	**L**		
TWO BENCHES					
F feet	1½"	2"	12"	2x4	4
G uprights	1½"	3½"	12"	2x4	8
H stretchers	1½"	2½"	60"	2x4	4
I top rails	1½"	3½"	10"	2x4	6
J bench tops	1½"	5½"	72"	2x6	4

Materials Key: choice of fir, pine, spruce, or redwood

Supplies: ¾" dowel stock, 2½" deck screws, 3" deck screws, 3¾" deck screws

bottom end of the uprights (G). See the photo *at right* for reference. Also, parts F and H are ripped to width from 2×4 stock. Finally, use the small heart pattern on the *opposite page* for the bench uprights (G).

FINISH THE PROJECTS

1 Sand the table and benches smooth (we used a palm sander with 100- and 150-grit sandpaper). Be sure to sand off any visible grade stamps marked on the boards.

2 Apply a finish. If you decide to paint your pieces, either an oil-based enamel or water-based latex will protect them. Regardless of your paint selection, be sure to apply a prime coat compatible with the top coat. Also, be sure to apply several coats to the porous end grain. For a

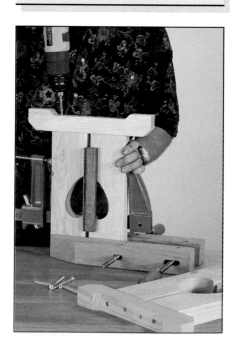

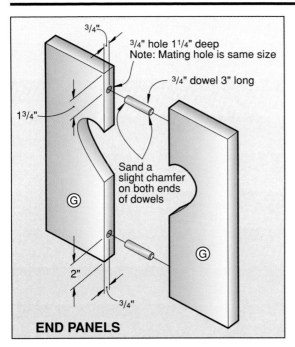

END PANELS

3/4"

3/4" hole 1 1/4" deep
Note: Mating hole is same size

3/4" dowel 3" long

1 3/4"

Sand a slight chamfer on both ends of dowels

G

G

2"

3/4"

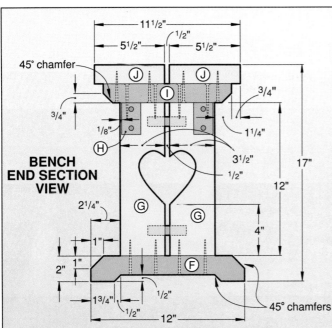

BENCH END SECTION VIEW

11 1/2"

5 1/2"   1/2"   5 1/2"

45° chamfer

J   J

I

3/4"

3/4"

1/8"

1 1/4"

H

3 1/2"

1/2"

17"

12"

G   G

4"

2 1/4"

1"

F

2"   1"

1 3/4"   1/2"

1/2"

12"

45° chamfers

Half-pattern for Ⓐ

Half-pattern for Ⓖ

FULL-SIZED HEART HALF-PATTERNS

more natural look, apply an exterior stain and several coats of spar varnish or other exterior sealer compatible with the stain. Be sure to choose an exterior finish with an ultra-violet inhibitor.

BUYING GUIDE

• **Bench, chair, and table plans.** For a 17×22" plan of the three accompanying pieces shown *at right*, send $9.95 to Country Comfy Trio WOOD PLANS, P.O. Box 9255, Dept. B-1, Des Moines, IA 50306 . The chair is featured on *page 34*.

Terrific Tambour Bookcase

We've seen a great many bookcases in our travels around the country, but none more good-looking and versatile than this one. And should you ever have to relocate, you'll find the modular pieces making up the bookcase a breeze to take apart and reassemble.

START BY FORMING THE EDGE JOINED OAK SHELVES

1 From ¾"-thick hardwood (we used oak), edge-join enough narrow stock to form three panels measuring 13" wide by 44" long. You'll use one panel for the upper shelf, another for the middle shelf, and the third for the lower shelf (all labeled A).

2 Rip and crosscut the three shelves to 12½" wide by 42" long. Sand both faces of each panel.

3 Mark a 2⅛" radius on each corner of each shelf. (See the Parts View drawing on the *opposite page* for reference.) Cut and sand the corners to shape. If you do this on the bandsaw, you'll need a helper to

support the long panels when making the corner cuts. Using a pencil, label one shelf UPPER, another MIDDLE, and the third, LOWER. Labeling the parts now helps ensure the correct machining to the proper face of each shelf later.

4 Drill four $^7/_{16}$" holes $^1/_2$" deep into the bottom side of the upper shelf, using the same centerpoints used to mark the $2^1/_8$"-radiused corners. Using a wide-blade screwdriver, drive a $^1/_4$"-threaded insert squarely into each of these holes until the notched end of the insert is flush with the bottom surface of the shelf where shown on the Threaded Insert detail accompanying the Exploded View drawing.

5 Switch bits, and drill $^5/_{16}$" holes through the middle shelf and lower shelf, again using the hole centerpoints you marked earlier. You'll slide the all-thread rod through these holes later when assembling the modules.

6 Make start and stop marks for the grooves ($2^1/_8$" from the ends) along the back edge on the bottom side of the upper shelf, both sides of the middle shelf, and top side of the lower shelf where dimensioned on the Parts View.

7 Fit your router with a $^1/_4$" straight bit and edge guide. If you don't have an edge guide, clamp a straightedge (a straight board will work) to the shelf. Rout a $^1/_4$" groove $^1/_4$" deep along the back edge of each

shelf where marked. You'll use these grooves to house the plywood backs (E).

8 Rout $^1/_4$" round-overs along the top and bottom edges of each shelf. Sand the shelves smooth.

NOW, LET'S CUT THE TAMBOUR STRIPS TO SIZE

1 To make four tambour support columns, you'll need 56 strips (B) per column, or a total of 224 strips for the four columns. (We cut about 240 strips, giving us extras to substitute for strips with defects.) To form the strips, plane thicker stock to $^7/_{16}$" thick. Crosscut the $^7/_{16}$"-thick stock to 29" long.

2 Set the fence on your tablesaw $^3/_{16}$" from the inside edge of the

continued

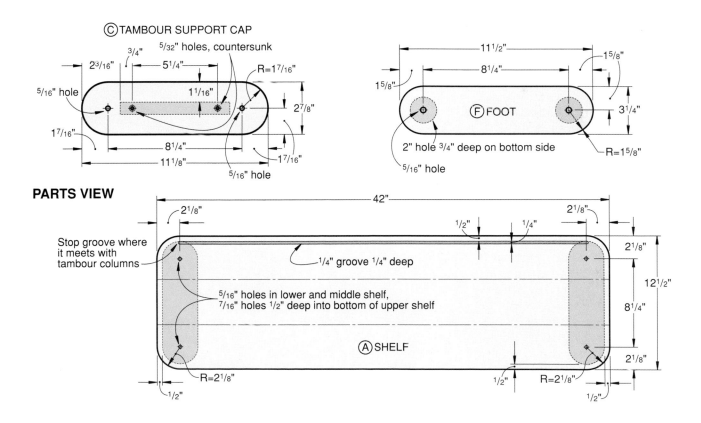

© TAMBOUR SUPPORT CAP

Ⓕ FOOT

2" hole $^3/_4$" deep on bottom side

$^5/_{16}$" hole

PARTS VIEW

Stop groove where it meets with tambour columns

$^1/_4$" groove $^1/_4$" deep

$^5/_{16}$" holes in lower and middle shelf, $^7/_{16}$" holes $^1/_2$" deep into bottom of upper shelf

Ⓐ SHELF

R=$2^1/_8$"

Terrific Tambour Bookcase

continued

blade (for smooth cuts, we used a 40-tooth, carbide-tipped saw blade). Rip 120 of the $\frac{3}{16} \times \frac{7}{16} \times 29$" strips (B).

3 Complete the B parts by crosscutting two 14"-long strips from each of the 29"-long strips.

GLUE AND CLAMP THE OAK STRIPS TO THE CANVAS

1 Build the Tambour Clamping Jig shown in Step 1 of the Making the Tambour drawing.

2 Cut four pieces of lightweight canvas to 12×30".

3 Starting at the end of the jig with the end cleat, position 56 of the $\frac{3}{16} \times \frac{7}{16} \times 14$" strips, best face down, on the jig where shown in Step 2 of the drawing. Place waxed paper under the first three and last three tambour strips. Use a pair of bar clamps to lightly clamp the strips against the end cleat to remove any gaps between the strips. Then, clamp a $\frac{1}{8} \times 2 \times 12$" stopblock in position to hold the strips tightly together.

4 Mark a line 1" in from the tambour-strip ends, and spread a thin, even coat of glue on all of the strips between the pair of marked lines. Place a piece of 12×30" lightweight canvas on the glued strips where shown in Photo A on *page* 46 and Step 2 of the drawing *at right*. Smooth out any wrinkles or bubbles in the canvas.

5 Cover the canvas with waxed paper. Next, clamp a piece of $\frac{3}{4} \times 12 \times 24\frac{1}{2}$" plywood to the top of this assembly to press the tambour strips uniformly against the canvas as shown in Photo B and Step 3 of the drawing. Leave the assembly clamped overnight. Remove the clamps and tambour, and repeat to form the remaining tambour assemblies.

6 Remove the tambour assembly from the jig, and sand the wood side of each tambour until the pieces
continued

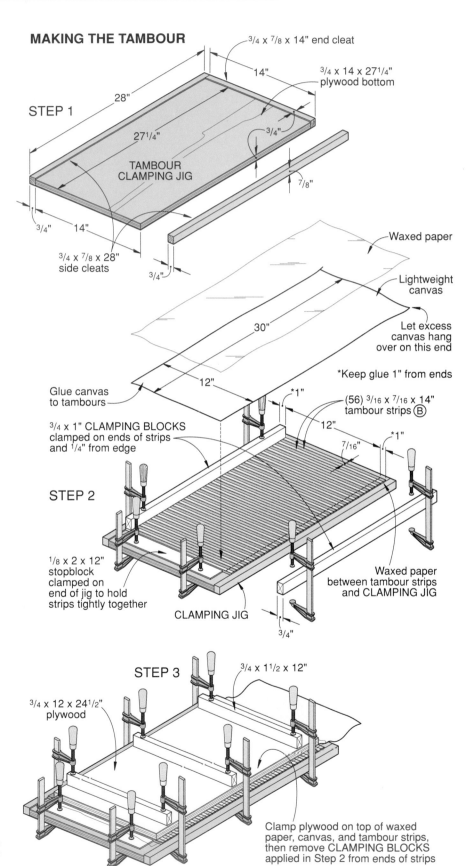

MAKING THE TAMBOUR

STEP 1

$\frac{3}{4} \times \frac{7}{8} \times 14$" end cleat
14"
$\frac{3}{4} \times 14 \times 27\frac{1}{4}$" plywood bottom
28"
$27\frac{1}{4}$"
TAMBOUR CLAMPING JIG
$\frac{3}{4}$"
$\frac{7}{8}$"
$\frac{3}{4}$"
14"
$\frac{3}{4} \times \frac{7}{8} \times 28$" side cleats
$\frac{3}{4}$"

Waxed paper
Lightweight canvas
Let excess canvas hang over on this end
*Keep glue 1" from ends
30"
12"
Glue canvas to tambours
*1"
(56) $\frac{3}{16} \times \frac{7}{16} \times 14$" tambour strips Ⓑ
$\frac{3}{4} \times 1$" CLAMPING BLOCKS clamped on ends of strips and $\frac{1}{4}$" from edge
12"
$\frac{7}{16}$"
*1"
STEP 2
$\frac{1}{8} \times 2 \times 12$" stopblock clamped on end of jig to hold strips tightly together
Waxed paper between tambour strips and CLAMPING JIG
CLAMPING JIG
$\frac{3}{4}$"

STEP 3
$\frac{3}{4} \times 1\frac{1}{2} \times 12$"
$\frac{3}{4} \times 12 \times 24\frac{1}{2}$" plywood
Clamp plywood on top of waxed paper, canvas, and tambour strips, then remove CLAMPING BLOCKS applied in Step 2 from ends of strips

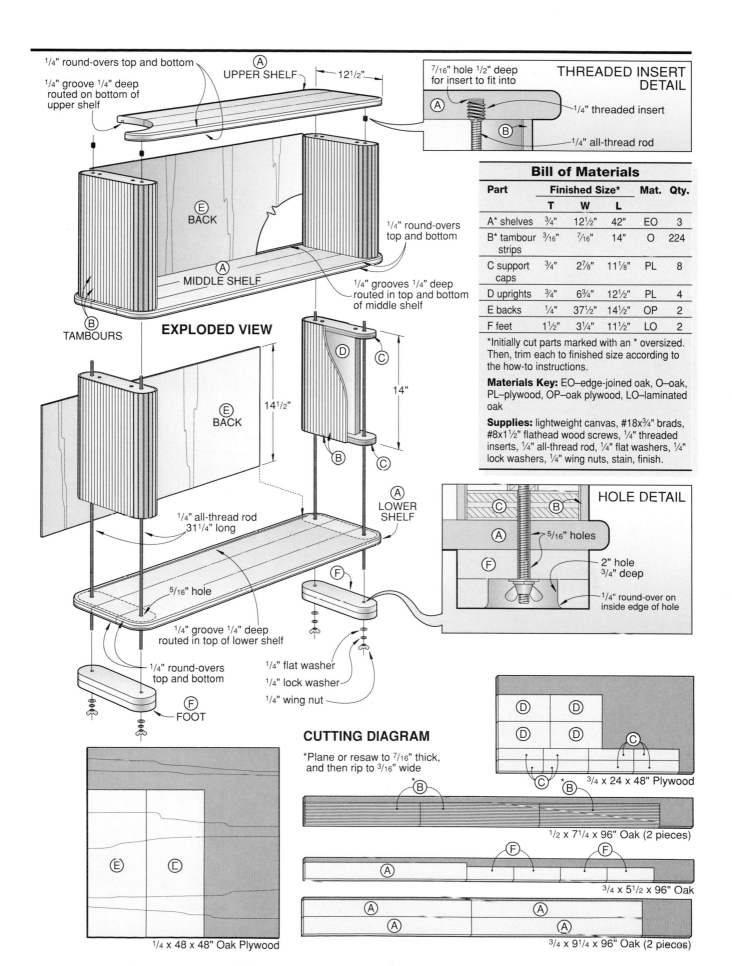

1/4" round-overs top and bottom

1/4" groove 1/4" deep routed on bottom of upper shelf

(A) UPPER SHELF — 12 1/2"

7/16" hole 1/2" deep for insert to fit into

THREADED INSERT DETAIL

(A) 1/4" threaded insert

(B) 1/4" all-thread rod

(E) BACK

1/4" round-overs top and bottom

(A) MIDDLE SHELF

1/4" grooves 1/4" deep routed in top and bottom of middle shelf

(B) TAMBOURS

EXPLODED VIEW

(D) (C)

14"

(E) BACK

14 1/2"

(B) (C)

1/4" all-thread rod 31 1/4" long

(A) LOWER SHELF

Bill of Materials

Part	Finished Size*			Mat.	Qty.
	T	W	L		
A* shelves	3/4"	12 1/2"	42"	EO	3
B* tambour strips	3/16"	7/16"	14"	O	224
C support caps	3/4"	2 7/8"	11 1/8"	PL	8
D uprights	3/4"	6 3/4"	12 1/2"	PL	4
E backs	1/4"	37 1/2"	14 1/2"	OP	2
F feet	1 1/2"	3 1/4"	11 1/2"	LO	2

*Initially cut parts marked with an * oversized. Then, trim each to finished size according to the how-to instructions.

Materials Key: EO–edge-joined oak, O–oak, PL–plywood, OP–oak plywood, LO–laminated oak

Supplies: lightweight canvas, #18x3/4" brads, #8x1 1/2" flathead wood screws, 1/4" threaded inserts, 1/4" all-thread rod, 1/4" flat washers, 1/4" lock washers, 1/4" wing nuts, stain, finish.

HOLE DETAIL

(C) (B)

(A) 5/16" holes

(F) 2" hole 3/4" deep

1/4" round-over on inside edge of hole

5/16" hole

1/4" groove 1/4" deep routed in top of lower shelf

(F)

1/4" round-overs top and bottom

1/4" flat washer
1/4" lock washer
1/4" wing nut

(F) FOOT

CUTTING DIAGRAM

*Plane or resaw to 7/16" thick, and then rip to 3/16" wide

(D) (D)

(D) (D)

(C)

(C) *

3/4 x 24 x 48" Plywood

* (B) (B)

1/2 x 7 1/4 x 96" Oak (2 pieces)

(F) (F)

(A)

3/4 x 5 1/2 x 96" Oak

(A) (A)

(A) (A)

3/4 x 9 1/4 x 96" Oak (2 pieces)

(E) (E)

1/4 x 48 x 48" Oak Plywood

Tambour Bookcase

continued

are flush with each other and all machining marks are gone. (We first belt-sanded the tambour with 80 and then 120-grit sandpaper. Next, we used a palm sander and 150- and 220-grit sandpaper to finish-sand the tambour.)

BUILD THE TAMBOUR FRAMES FOR A SUPPORTING ROLE

1 To build the tambour support frames, cut the support caps (C) to the shape shown on the Parts View drawing. Cut the uprights (D) to size.

2 Using the Parts View drawing for reference, mark the centerpoints and drill four holes in each top and bottom support cap (C).

3 Glue and screw the supports together in the configuration shown on the Tambour Support Frame drawing. (We used a framing square to check that the edges of the top and bottom caps (C) were perfectly aligned.)

4 To wrap the tambour around the frames, drill 1/16" holes 3/8" from the ends of the first two tambour strips where shown on Step 1 of the Assembling the Columns drawing on the *opposite page.*

5 Spread glue on the mating edges and wrap the tambour tightly around the frame where shown in Step 1 of the drawing. Start the first strip centered on the radiused ends where shown.

6 After gluing and wrapping the tambour completely around the support frame, clamp a pair of clamp blocks to the top and bottom of the assembly where shown in Step 2 of the drawing. The clamp blocks hold the ends of the tambour strips flush while the glue dries. Repeat the process to form each column. Drive brads into the last two tambour strips in the wrap to hold them firmly in place.

Clamp the oak strips to the clamping jig good face down. Spread glue between the clamping blocks, and adhere a piece of lightweight canvas to the strips.

Place waxed paper on top of the canvas, and then position and clamp a piece of plywood on top of the waxed paper to firmly secure the canvas to the oak strips.

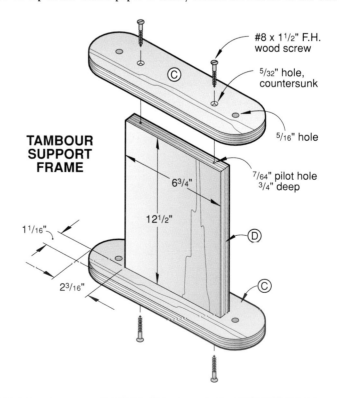

#8 x 1½" F.H. wood screw

5/32" hole, countersunk

5/16" hole

7/64" pilot hole 3/4" deep

TAMBOUR SUPPORT FRAME

6¾"

12½"

1 1/16"

2 3/16"

C

D

C

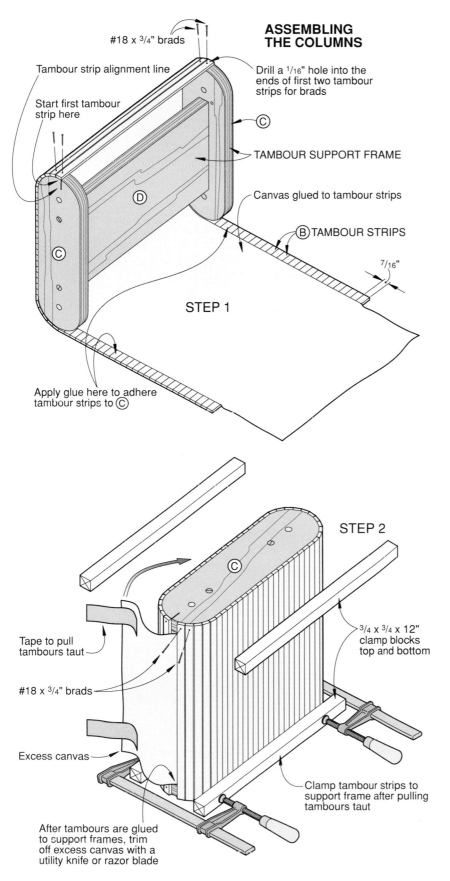

ASSEMBLING THE COLUMNS

#18 x 3/4" brads

Tambour strip alignment line

Start first tambour strip here

Drill a 1/16" hole into the ends of first two tambour strips for brads

TAMBOUR SUPPORT FRAME

Canvas glued to tambour strips

B TAMBOUR STRIPS

7/16"

STEP 1

Apply glue here to adhere tambour strips to C

STEP 2

Tape to pull tambours taut

#18 x 3/4" brads

Excess canvas

3/4 x 3/4 x 12" clamp blocks top and bottom

Clamp tambour strips to support frame after pulling tambours taut

After tambours are glued to support frames, trim off excess canvas with a utility knife or razor blade

7 Using a hobby knife, trim the excess canvas from each column. Set the brads, and fill the holes with putty. Sand smooth.

8 Belt-sand the top and bottom surfaces of the column flush. Hand-sand a slight round-over on the exposed edges of the tambour pieces where they wrap around the frame radiused ends.

CUT THE BACKS, AND PUT THE PIECES TOGETHER

1 From 1/4" oak plywood, cut the backs (E) to size. Notice that the grain runs up and down on the backs where shown on the Exploded View drawing.

2 Cut two pieces of 3/4"-thick stock for each foot (F) to the shape shown on the Parts View drawing. If you don't have a 2" Forstner bit to form the holes in the bottom pieces, drill blade start holes and cut the openings with a scrollsaw or jigsaw. Glue the pieces face-to-face with the edges and ends flush. Sand the edges smooth.

3 Drill 5/16" holes through each foot where shown on the Parts View drawing.

4 Finish-sand all of the pieces. Stain the components and apply a clear finish.

5 Using a hacksaw, cut two 31 1/4"-long pieces of 1/4" all-thread rod. Assemble the bookcase in the configuration shown on the Exploded View drawing. To prevent scratching the floor, make sure that the bottom of the all-thread is above the bottom surface of the feet where shown on the Hole detail accompanying the Exploded View drawing. Trim the all-thread rod more if necessary.

Masterpiece in Pine

Build this versatile cabinet for your country-loving home

Need a handsome country entertainment center to hold TV and stereo equipment? Or, how about a traditional wardrobe for your bedroom? This sharp-looking cabinet, made from pine, can perform either role—and do it in style.

LET'S START WITH THE CARCASE ASSEMBLY

1 Edge-join narrower stock to size to form the cabinet sides (A), bottom (B), and top (C) to the sizes listed in the Bill of Materials plus ½" in width and 1" in length.

2 Lay out the panels for the best grain match. Mark the edges that will receive the splines. See the Carcase Assembly drawing and Spline detail for reference.

3 Using a ¼" slotting cutter, rout ¼" slots, centered, along the marked edges of the panels. Stop the slots 2" from the ends for the top panel (C). See the Spline detail for reference. The slots in the Side and bottom panels go all the way through.

4 From ¼" stock (we used plywood), cut $^{15}/_{16}$"-wide splines to the lengths needed to fit the routed slots. Cut or sand the ends of the top-panel splines to shape.

5 Glue, spline, and clamp the side panels (A), bottom panel (B), and top panel (C) together, checking that the panels remain flat. Remove excess glue.

6 Rip and crosscut the panels (A, B, C) to the finished sizes listed in the Bill of Materials.

7 To fit the shelf standards later, mark the locations, and cut a pair of ⅝" grooves ¼" deep 2½" from the front and back edges in the side panels.

IT'S TIME TO ASSEMBLE THE CARCASE

1 Rip and crosscut the bottom front rail (D) and top front rail (E) to size from ¾"-thick pine stock.

2 Cut the cleats (F, G, H) to size. Drill and countersink the mounting holes through the cleats.

3 Position and square the side cleats (F) on the cabinet sides (A) where located on the Carcase Assembly drawing. Screw the cleats in place. Position and screw the top cleats (G) in place.

4 Dry-clamp the bottom and top panels (B, C) to the side panels. The back edges of both panels are flush with the back edge of the sides. The top panel overhangs both side panels by 3". Check for square and screw (no glue) the panels together.

5 Dry-clamp the rails (D, E) in place, drill countersunk screw holes, and glue and screw these pieces to the carcase.

6 Rout a ⅜" rabbet ¼" deep along the back inside edge of the assembled cabinet. Chisel the round-routed corners square.

7 Measure the rabbeted opening, and cut the back (I) to size from ¼" hardboard. If you plan to use the cabinet for electronic components, bore ventilation and wire access holes. See the Carcase Assembly drawing for reference.

NEXT, ADD THE DECORATIVE PINE MOLDING

1 Cut the bottom trim pieces (J, K, L, M) to size plus 1" in length.

2 Rout a ½" cove along the top front edge of the inner trim pieces (J, K). See the Lower Molding detail accompanying the Exploded View drawing for reference. Miter-cut the pieces and screw them (no glue) to the cabinet bottom.

3 Rout a ½" bead along the top front edge of molding pieces (L, M). Miter-cut these pieces and glue them to the cabinet.

continued

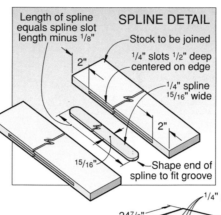

SPLINE DETAIL

Length of spline equals spline slot length minus ⅛"

Stock to be joined

¼" slots ½" deep centered on edge

¼" spline 15/16" wide

2"

2"

15/16"

Shape end of spline to fit groove

Adjustable shelves make this handsome project versatile enough to fulfill a number of storage needs.

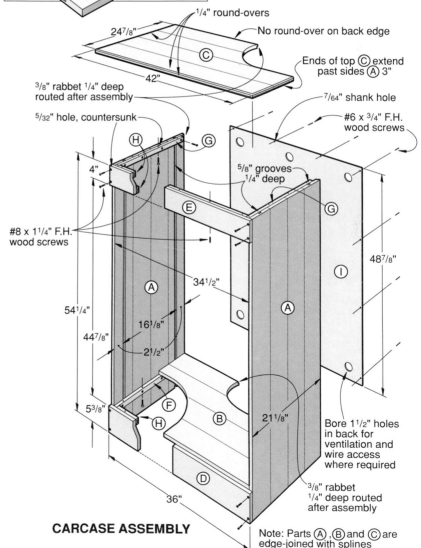

¼" round-overs

No round-over on back edge

24⅞"

C

42"

Ends of top C extend past sides A 3"

⅜" rabbet ¼" deep routed after assembly

7/64" shank hole

5/32" hole, countersunk

#6 x ¾" F.H. wood screws

5/8" grooves ¼" deep

H

G

4"

E

G

#8 x 1¼" F.H. wood screws

I

A

34½"

A

48⅞"

54¼"

16⅛"

44⅞"

2½"

F

B

21⅛"

5⅜"

H

3/8" rabbet ¼" deep routed after assembly

D

Bore 1½" holes in back for ventilation and wire access where required

36"

CARCASE ASSEMBLY

Note: Parts A, B and C are edge-joined with splines

Masterpiece in Pine

continued

4 Cut the top trim pieces (N, O) to size, and screw (no glue) them to the top of the cabinet.

5 Cut dentil mold pieces (P, Q) to size plus 2" in length. Construct the jig shown *below right*. The *center* of the ⅛" kerf in the miter extension should be ½" from the center of the indexing pin.

6 Make one kerfing cut across the end of one of the dentil mold strips. Next, position this kerf on the indexing pin, and make the second cut. Repeat the process as shown in the photo *below* to make all the evenly spaced kerfs. Repeat for each piece.

Use a wooden miter-gauge extension with a guide for cutting the dentil mold.

7 Miter-cut the dentil mold pieces (P, Q), and glue them to the cabinet where shown on the Exploded View drawing.

8 To add the pine crown molding (R, S), we purchased an 8' piece at a local homecenter. Select a straight, flat 2×4, and glue it to the back face of the crown molding where shown on Step 1 of the Ripping the Crown Molding drawing. Now, rip the edges of the lamination where shown in Steps 1 and 2 of the drawing.

9 Measure and miter-cut the front piece (S) to length and glue it to the top front of the cabinet. Repeat for the side pieces (R).

CUTTING DIAGRAM

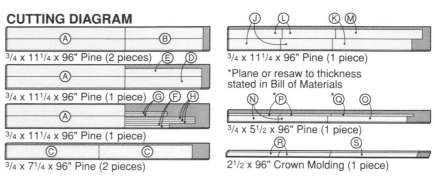

¾ x 11¼ x 96" Pine (2 pieces)

¾ x 11¼ x 96" Pine (1 piece)

¾ x 11¼ x 96" Pine (1 piece)

¾ x 7¼ x 96" Pine (2 pieces)

¾ x 11¼ x 96" Pine (1 piece)

*Plane or resaw to thickness stated in Bill of Materials

¾ x 5½ x 96" Pine (1 piece)

2½ x 96" Crown Molding (1 piece)

Bill of Materials

Part	Finished Size*			Mat.	Qty.
	T	W	L		
CABINET CARCASE					
A* sides	¾"	21⅛"	54¼"	EP	2
B* bottom	¾"	21⅛"	34½"	EP	1
C* top	¾"	24⅞"	42"	EP	1
D bottom rail	¾"	5⅜"	36"	P	1
E top rail	¾"	4"	36"	P	1
F cleats	¾"	1"	21⅛"	P	2
G cleats	¾"	1"	20"	P	2
H cleats	¾"	1"	33"	P	3
I back	¼"	35¼"	48⅞"	H	1
MOLDING					
J* btm. sides	¾"	5¼"	22⅝"	P	2
K* btm. front	¾"	5¼"	37½"	P	1
L* btm. sides	¾"	4⅝"	23⅜"	P	2
M* btm. front	¾"	4⅝"	39"	P	1
N top sides	¾"	2¾"	22⅝"	P	2
O top front	¾"	2¾"	37½"	P	1
P* dentil mold	½"	1"	22⅜"	P	2
Q* dentil mold	½"	1"	37"	P	1
R* crown mold	1⅞"	2½"	24½"	P	2
S* crown mold	1⅞"	2½"	41¼"	P	1

Part	Finished Size*			Mat.	Qty.
	T	W	L		
DOORS					
T stiles	¾"	2½"	44¾"	P	4
U top rails	¾"	2½"	14"	P	2
V btm. rails	¾"	3¾"	14"	P	2
W* panels	¾"	13⅞"	39⁷⁄₁₆"	EP	2
SHELVES					
X* shelves	¾"	18"	34⅜"	EP	3
Y* fronts	¾"	1½"	34⅜"	P	3

*Initially cut parts marked with an * oversized. Then, trim each to finished size according to the how-to instructions.

Materials Key: EP–edge-joined pine, P–pine, H–hardboard.

Supplies: #6x½" flathead wood screws, #6x¾" flathead wood screws, #8x1¼" flathead wood screws, 8' pine crown molding, cabinet top and bottom pivot hinges (Stanley catalog no. 332), cabinet middle pivot hinges (Stanley catalog no. 335), pulls (Amerock BP76271-R1), roller catches (Amerock BP9745-3), 4–⅝" brown or walnut finished shelf standards, wood conditioner, stain, clear finish.

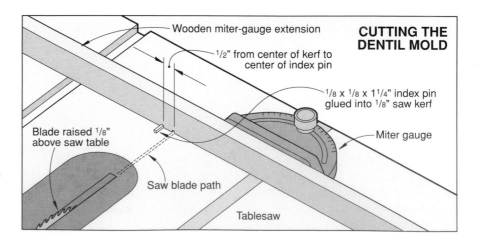

Wooden miter-gauge extension

½" from center of kerf to center of index pin

⅛ x ⅛ x 1¼" index pin glued into ⅛" saw kerf

CUTTING THE DENTIL MOLD

Blade raised ⅛" above saw table

Miter gauge

Saw blade path

Tablesaw

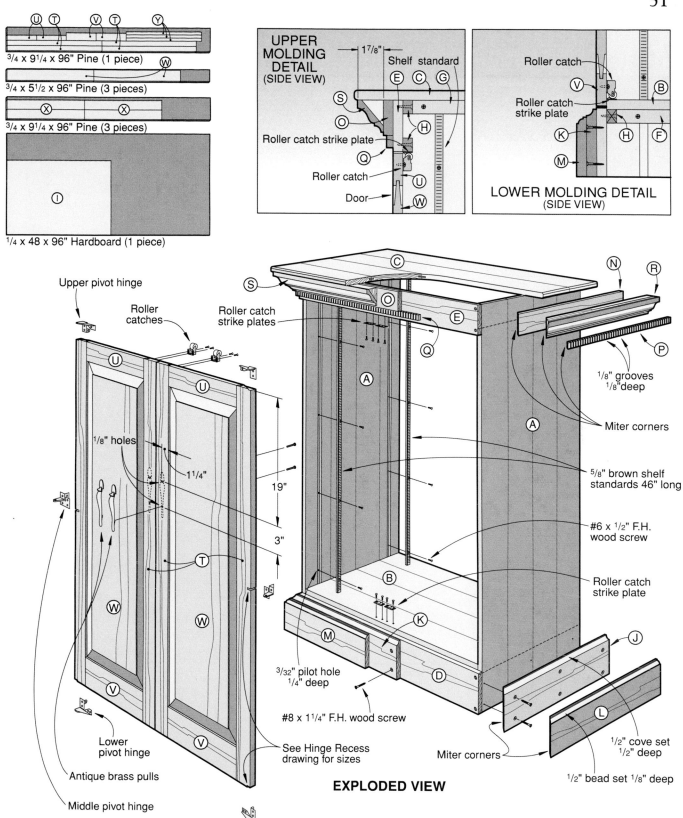

UPPER MOLDING DETAIL (SIDE VIEW)

1⁷⁄₈"

Shelf standard

Roller catch strike plate

Roller catch

Door

LOWER MOLDING DETAIL (SIDE VIEW)

Roller catch

Roller catch strike plate

¾ x 9¼ x 96" Pine (1 piece)

¾ x 5½ x 96" Pine (3 pieces)

¾ x 9¼ x 96" Pine (3 pieces)

¼ x 48 x 96" Hardboard (1 piece)

Upper pivot hinge

Roller catches

Roller catch strike plates

⅛" holes

1¼"

19"

3"

¼" holes

Lower pivot hinge

Antique brass pulls

Middle pivot hinge

3/32" pilot hole ¼" deep

#8 x 1¼" F.H. wood screw

See Hinge Recess drawing for sizes

Miter corners

⅛" grooves ⅛" deep

Miter corners

⅝" brown shelf standards 46" long

#6 x ½" F.H. wood screw

Roller catch strike plate

½" cove set ½" deep

½" bead set ⅛" deep

EXPLODED VIEW

NOW, CONSTRUCT A PAIR OF DOORS TO CLOSE THE CASE

1 Cut the door stiles (T), top rails (U), and bottom rails (V) to size.

2 Attach a ¼" dado blade to your tablesaw, and raise it ½" above the surface of the saw table. Cut a ¼" groove, centered from side to side, along one edge of each part. (We test-cut the groove in scrap stock the same thickness as the stiles and rails to ensure it was centered.) See the Door drawing and accompanying Tenon detail for reference.

continued

Masterpiece in Pine

continued

3 To cut the stub tenons on the ends of the rails, use a miter gauge with an attached wood extension for support, and cut a ½" rabbet ¼" deep on both faces of each end of each rail.

4 For stability and flatness, edge-join narrower, straight-grained stock to form the two door panels (W) to the size listed in the Bill of Materials plus 1" in length and width. Later, cut the door panels to finished size.

5 To cut the raised panels, tilt your tablesaw blade 5° from vertical, and elevate the blade 2" above the saw table. Position the rip fence, test-cut scrap stock to create the same shape shown on the Raised-Panel detail, and cut each edge of each panel. Sand the cut areas to remove saw marks.

6 Test-fit the door pieces, and trim if necessary. Then, finish-sand the panels and stain them. (For a more even stain application on the pine, we applied Minwax Wood Conditioner before applying the stain.) Then, glue and clamp each door, checking for square. Rest the clamped assemblies on a flat surface.

ADD THE PIVOT HINGES TO THE DOORS AND CARCASE

1 Mark the upper and lower hinge recesses where shown on the Top and Bottom Hinge Recess drawings. Then, mark the locations for the middle notches where shown on the Middle Hinge Notch drawing. Follow the instructions supplied with the pivot hinges to form the top, bottom, and middle hinge recesses and notches.

2 Mount the hinges to the doors, and fasten the hinges to the cabinet. Mount the roller catches to

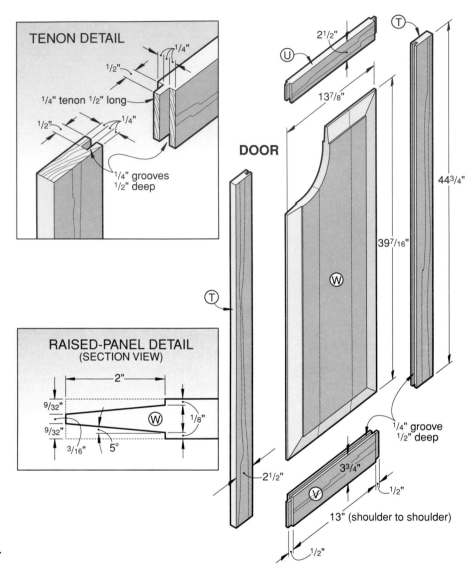

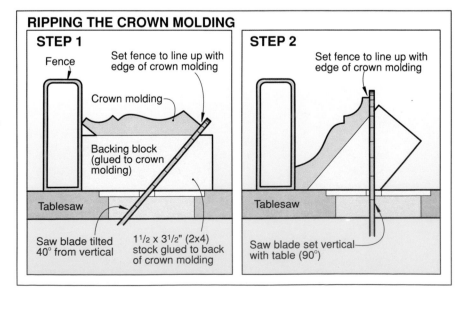

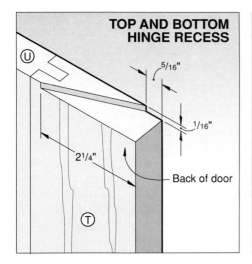

TOP AND BOTTOM HINGE RECESS

U

5/16"

1/16"

2 1/4"

T

← Back of door

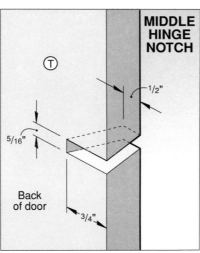

MIDDLE HINGE NOTCH

T

1/2"

5/16"

Back of door

3/4"

the doors and cabinet. Drill the holes for the pulls. Attach the pulls and catches.

3 Crosscut the shelf standards, and install them in the cabinet.

ADD THE SHELVES AND FINISH

Note: *The shelves are dimensioned to allow a little over an inch of clearance between the back edge of the shelf and the front edge of the back piece (I). The gap allows passage for the cords from your electrical components to the wire-access holes.*

1 Using the same procedure you used to form the panels for the side, bottom, and top panels, edge-join and spline enough pine stock for the number of shelves (X) that you'll need.

2 Cut the shelves to fit between the shelf standards.

3 Cut the shelf fronts (Y) to size plus 1" in length. Cut a 3/4" rabbet 1/4" deep along the back edge of each. See the Shelf Recess detail for reference. Glue and clamp a front to each shelf.

4 Rout 1/4" beads along the front edge of each shelf front. Trim the ends of the shelf fronts flush with the ends of each shelf.

5 Fit the clips on the shelf standards. Form clip recesses on the bottom of each shelf where dimensioned on the Shelf drawing and Shelf Recess detail.

6 Remove all the hardware from the cabinet and doors. Remove the back (I) from the cabinet. Finish-sand all the pieces. Apply wood conditioner, and then stain and finish. (We used Minwax Wood Conditioner, Puritan Pine stain, and satin polyurethane.)

7 Reattach the hardware and hang the doors. Screw the cabinet back in place.

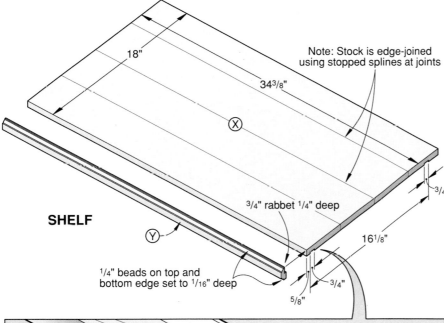

18"

34 3/8"

Note: Stock is edge-joined using stopped splines at joints

X

3/4"

SHELF

Y

3/4" rabbet 1/4" deep

16 1/8"

3/4"

5/8"

1/4" beads on top and bottom edge set to 1/16" deep

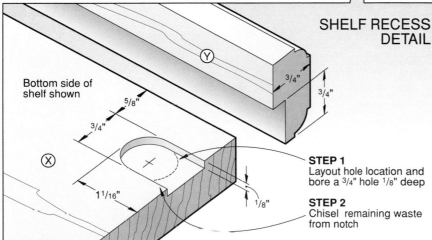

SHELF RECESS DETAIL

Bottom side of shelf shown

Y

3/4"

3/4"

5/8"

3/4"

X

1 1/16"

1/8"

STEP 1
Layout hole location and bore a 3/4" hole 1/8" deep

STEP 2
Chisel remaining waste from notch

Stylish Storage

A hardwood hideaway for your CDs, videotapes, and cassettes

Looking for a way to make some order out of those ever-growing stacks of cassettes, tapes, and CDs at your house? We can help. Our solid walnut cabinet, which features pull-out shelves and solid brass hardware, will meet the challenge—and look good doing it.

START WITH THE BASIC CABINET

1 Cut the cabinet sides (A) and top and bottom (B) ¼" wider and 1" longer than the size listed in the Bill of Materials. (We edge-joined narrower stock to form the wide panels.) Later, scrape off the excess glue and trim the four panels to finished size.

2 Cut the rabbets and dadoes in the cabinet sides (A) where shown and to the sizes shown on the Basic Cabinet Assembly drawing. Don't forget to machine a ¼" rabbet ½" deep along the back inside edge of the side panels.

3 Cut the cabinet front stiles (C) to size plus 1" in length.

4 With the surfaces flush and an even overhang on both ends, glue and clamp a stile (C) to the front edge of each cabinet side (A). Trim the ends of the stiles flush with the ends of the side panels. Sand the side panels smooth.

5 Cut a ½" notch 1" deep in the front corners of each top and bottom panel (B).

6 Glue and clamp the top and bottom (B) between the side panels. Check for square and that the front edges are flush. The back edges of the top and bottom panels should sit in ¼", leaving room for the back (D) where shown on the Back detail accompanying the Basic Cabinet Assembly drawing.

7 Measure the opening, and cut the back (D) to size from ¼" tempered hardboard. Fit the back in place, and drill mounting holes to the sizes listed on the Basic Cabinet Assembly drawing. Do not attach the back yet.

continued

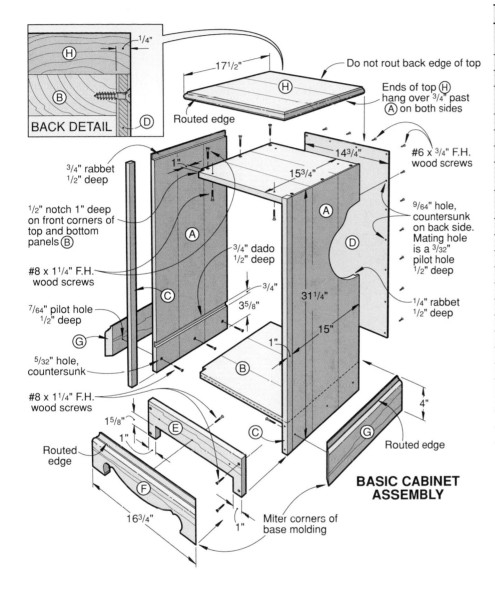

BACK DETAIL

1/4"

Do not rout back edge of top

17½"

Ends of top (H) hang over ¾" past (A) on both sides

Routed edge

14¾"

15¾"

#6 x ¾" F.H. wood screws

¾" rabbet ½" deep

1"

9/64" hole, countersunk on back side. Mating hole is a 3/32" pilot hole ½" deep

½" notch 1" deep on front corners of top and bottom panels (B)

¾" dado ½" deep

¾"

31¼"

¼" rabbet ½" deep

#8 x 1¼" F.H. wood screws

3⅝"

15"

7/64" pilot hole ½" deep

1"

5/32" hole, countersunk

#8 x 1¼" F.H. wood screws

1⅝"

1"

Routed edge

Routed edge

4"

16¾"

1"

Miter corners of base molding

BASIC CABINET ASSEMBLY

Bill of Materials

Part	Finished Size*			Mat.	Qty.
	T	W	L		
BASIC CABINET					
A* sides	¾"	15"	31¼"	EW	2
B* top & bottom	¾"	15¾"	14¾"	EW	2
C* front stiles	¾"	1"	31¼"	W	2
D back	¼"	14¾"	27⅝"	TH	1
E support	¾"	4"	15¼"	W	1
F* front base molding	¾"	4"	16¾"	W	1
G* side base molding	¾"	4"	17½"	W	2
H* top	¾"	17½"	16¾"	EW	1
DOOR					
I stiles	¾"	1¾"	27"	W	2
J lower rail	¾"	2¼"	11¾"	W	1
K upper rail	¾"	1¾"	11¾"	W	1
L* panel	½"	11¹¹⁄₁₆"	22¹⁵⁄₁₆"	EW	1
M* stops	¼"	½"	23"	W	2
N* stops	¼"	½"	11¾"	W	2
O* stops	¼"	¼"	23"	W	2
P* stops	¼"	¼"	11¾"	W	2
SHELVES					
Q guides	½"	1⅛"	14½"	W	8
R fronts	¾"	2⅜"	13½"	W	4
S bottoms	¼"	13⅛"	14¾"	TH	4

*Initially cut parts marked with an * oversized. Then, trim each to finished size according to the how-to instructions.

Materials Key: EW–edge-joined walnut, W–walnut, TH–tempered hardboard

Supplies: #6x½" flathead brass wood screws, #6x¾" flathead wood screws, #8x1" flathead wood screws, #8x1¼" flathead wood screws, ⅜x1½" dowel pins (those with glue grooves), double-faced tape, #8x½" flathead brass wood screws, #17x¾" brads, clear finish.

CUTTING DIAGRAM

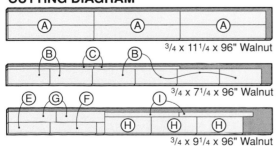

(A) (A) (A)
¾ x 11¼ x 96" Walnut

(B) (C) (B)
¾ x 7¼ x 96" Walnut

(E) (G) (F) (I)
(H) (H) (H)
¾ x 9¼ x 96" Walnut

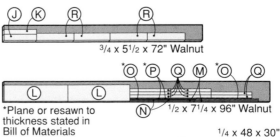

(J) (K) (R) (R)
¾ x 5½ x 72" Walnut

(O) (P) (Q) (M) (O) (Q)
(L) (L)
(N)
½ x 7¼ x 96" Walnut

*Plane or resawn to thickness stated in Bill of Materials

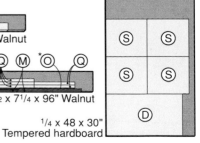

(S) (S)
(S) (S)
(D)
¼ x 48 x 30" Tempered hardboard

Stylish Storage

continued

NOW, ADD THE BASE MOLDING AND CABINET TOP

1 Using the dimensions on the Basic Cabinet Assembly drawing, cut the front base-molding support (E) to shape. Drill the holes and screw it to the cabinet.

2 For base molding parts (F, G), cut a piece of ³/₄" walnut to 4×54". Following the two-step Routing the Base Molding drawing *below*, rout along one edge of the 54"-long piece.

3 Miter-cut the front (F) and then the sides (G) to length. Transfer the full-sized half-pattern to the front (F) twice to mark the curves along the bottom edge. Bandsaw the curved section to shape, and drum-sand smooth. Drill countersunk holes through the inside surface of A and E and into the inside face of F and G. Screw the pieces to the cabinet.

4 Edge-join enough stock for the cabinet top (H). Later, trim the top to size.

5 Following the three-step Routing the Top drawing *below*, rout the decorative edge along the front and side edges (not the back) of the cabinet top. Sand the routed edges smooth. With the back edges flush and the top centered from side-to-side, screw the top to the cabinet.

THE RAISED-PANEL DOOR COMES NEXT

1 Cut the two door stiles (I) and rails (J, K) to size.

2 Using the dimensions on the Door drawing and accompanying Section View Detail, mark the dowel-hole locations and drill the holes. (We used a doweling jig.)

3 Using dowel pins, glue and dowel the door frame together.

4 Edge-join ½"-thick stock to form the door panel (L) ¼" wider and ½" longer than the size listed in the Bill of Materials. (We book-matched two pieces of highly figured walnut for the door panel.)

5 Measure the opening and cut the door panel (L) ¹/₁₆" shorter and narrower than the opening.

6 Using the two-step drawing titled Making the Raised Panel *below*, machine the edges of the door panel.

7 Mark the locations and form the three ¹/₁₆"-deep hinge mortises in the right-hand door stile where dimensioned on the Final Assembly drawing. Center the door top to bottom on the front of the cabinet and clamp it in place. Mark lines to transfer the hinge locations from the door to the cabinet, remove the door, and form the mortises where marked.

8 Attach the catches to the cabinet where shown on the Final Assembly drawing.

continued

ROUTING THE BASE MOLDING

STEP 1

F and G
Fence
½"
¼"
Router table
½"-dia. round-nose bit

STEP 2

Fence
¼"
F and G
5/16"
Router table   ¼" round-over bit

ROUTING THE TOP

STEP 1

Fence
3/8"
Bottom face of H
3/16"
Router table   3/4"-dia. round-nose bit

STEP 2

Fence
1/4"
Bottom face of H
3/16"
Router table   1/4" round-over bit

STEP 3

Fence
1/4"
Top face of H
1/16"
Router table   1/4" round-over bit

MAKING THE RAISED PANEL

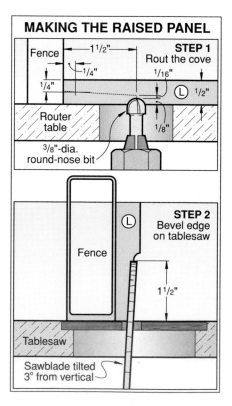

STEP 1
Rout the cove

Fence   1 1/2"
1/4"
1/16"
1/4"   L   1/2"
Router table   1/8"
3/8"-dia. round-nose bit

STEP 2
Bevel edge on tablesaw

L
Fence
1 1/2"
Tablesaw
Sawblade tilted 3° from vertical

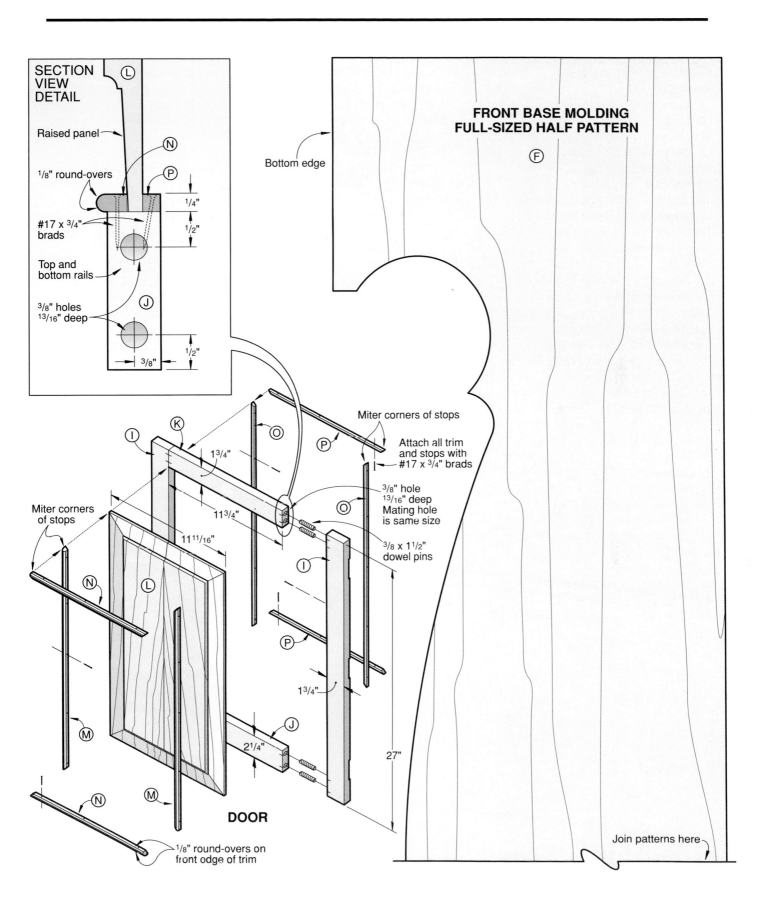

SECTION VIEW DETAIL

Ⓛ

Raised panel

Ⓝ

Ⓟ

1/8" round-overs

1/4"

1/2"

#17 x 3/4" brads

Top and bottom rails

Ⓙ

3/8" holes
13/16" deep

1/2"

3/8"

**FRONT BASE MOLDING
FULL-SIZED HALF PATTERN**

Ⓕ

Bottom edge

Miter corners of stops

Attach all trim and stops with #17 x 3/4" brads

Ⓘ Ⓚ Ⓞ Ⓟ

1 3/4"

11 3/4"

11 11/16"

3/8" hole
13/16" deep
Mating hole is same size

Ⓞ

3/8 x 1 1/2" dowel pins

Ⓘ

Miter corners of stops

Ⓝ

Ⓛ

Ⓟ

1 3/4"

Ⓜ

27"

Ⓙ

2 1/4"

DOOR

Ⓝ

Ⓜ

1/8" round-overs on front odge of trim

Join patterns here

Stylish Storage

continued

CUT AND INSTALL THE PANEL STOPS

1 Plane or resaw one piece of 3×36" walnut to ¼" thick. Rout ⅛" round-overs along both edges of the ¼" strip. See the Section View Detail accompanying the Door drawing for the routed profiles.

2 Rip a ½"-wide strip from each edge of the 3"-wide board for stops (M, N). Then, rip two ¼"-wide strips from the remaining stock for stops (O, P).

3 Miter-cut one stop M and one N from each ¼ × ½ × 36" strip. Miter-cut the back stops (O, P) to length from the ¼ × ¼ × 36" strips.

4 Make a bit to drill the pilot holes through the walnut stops for the nails. To do this, snip the head off a 4d finish nail.

5 Chuck the headless nail into your portable drill. Turn the drill on. By holding the rotating nail against a drum sander or disc sander, reduce the diameter of the nail to the same diameter as the brad. Sharpen the point. Use the drill and "pilot bit" to drill the pilot holes through the panel stops. As shown in the Section View Detail, we angled the pilot holes in the stops.

6 Build the panel stop positioning jig as shown *below left*. Use the jig to position the front stops (M, N), and then drive the brads as shown in the photo *below center*. Set the brads and fill the holes (we used FIX walnut putty). Set the back stops (O, P) aside for now; you'll add them later.

AND NOW FOR THE PULL-OUT SHELVES

Note: *The location of the drawer guides (Q) against the inside face of the cabinet sides (A) will depend on the type of tape holders you use. Test-fit the guides (we temporarily held them in place with double-faced tape) before screwing them in position.*

As shown in the opening inset photo and Final Assembly drawing, we used two pair of audio cassette holders on the top shelf, two pair of compact disc holders on the second shelf, and one pair of video cassette holders on the bottom two shelves. See the Buying Guide for our source of plastic holders and hardware.

1 Cut the guides (Q) to the size listed in the Bill of Materials. Cut a ¼" groove ¼" deep in each guide where shown on the Guide detail on the Final Assembly drawing.

2 Drill countersunk mounting holes through each guide, and screw them in place. (As shown in the photo *below right*, we used a spacer to ensure accurate spacing between the guides from one side of the cabinet to the other.)

3 Cut the shelf fronts (R) to size. Rout the edges of each shelf front as indicated on the Shelf Section View Detail accompanying the Final Assembly drawing.

4 Cut the shelf bottoms (S) to size. Check the fit of the bottoms into the shelf guides in the cabinet. The shelves should slide back and forth easily.

5 Glue the shelf bottoms into the ¼" grooves in the shelf fronts, checking that the shelves are square to the fronts.

ADD THE FINISH, AND POSITION THE TAPE HOLDERS

1 Remove the hardware, and finish-sand all the parts.

2 Apply finish to the cabinet, door, door panel, back stops, and shelf fronts. Do not apply finish to the hardboard shelf bottoms (S). (We applied several coats of Watco Dark Walnut Oil Finish.)

3 Place the door panel in the door and secure the back stops (O, P). Set the nails, putty the holes, and wipe off any excess. Drill the mounting hole in the door for the pull. See the Final Assembly drawing for reference.

4 Adhere the tape holders to the top surface of the shelves.

5 Screw the back (D) onto the cabinet. Attach the hinges, pull, and catches.

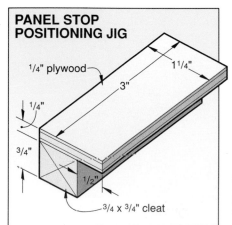

PANEL STOP POSITIONING JIG

¼" plywood

1¼"

3"

¼"

¾"

½"

¾ × ¾" cleat

BUYING GUIDE

• **Hardware and tape holders.** For current prices of two pair $1\frac{3}{4} \times 2$" solid-brass cabinet hinges (32268), two $\frac{1}{4} \times 1\frac{3}{4}$" brass ball catches and strikes (28613), one brass pendant pull (35527), black plastic compact disc holders (30536), video cassette holders (30528), or audio cassette holders (30510), contact The Woodworkers' Store, 21801 Industrial Blvd., Rogers, MN 55374–9514. Or call 1-800-260-9663 to order.

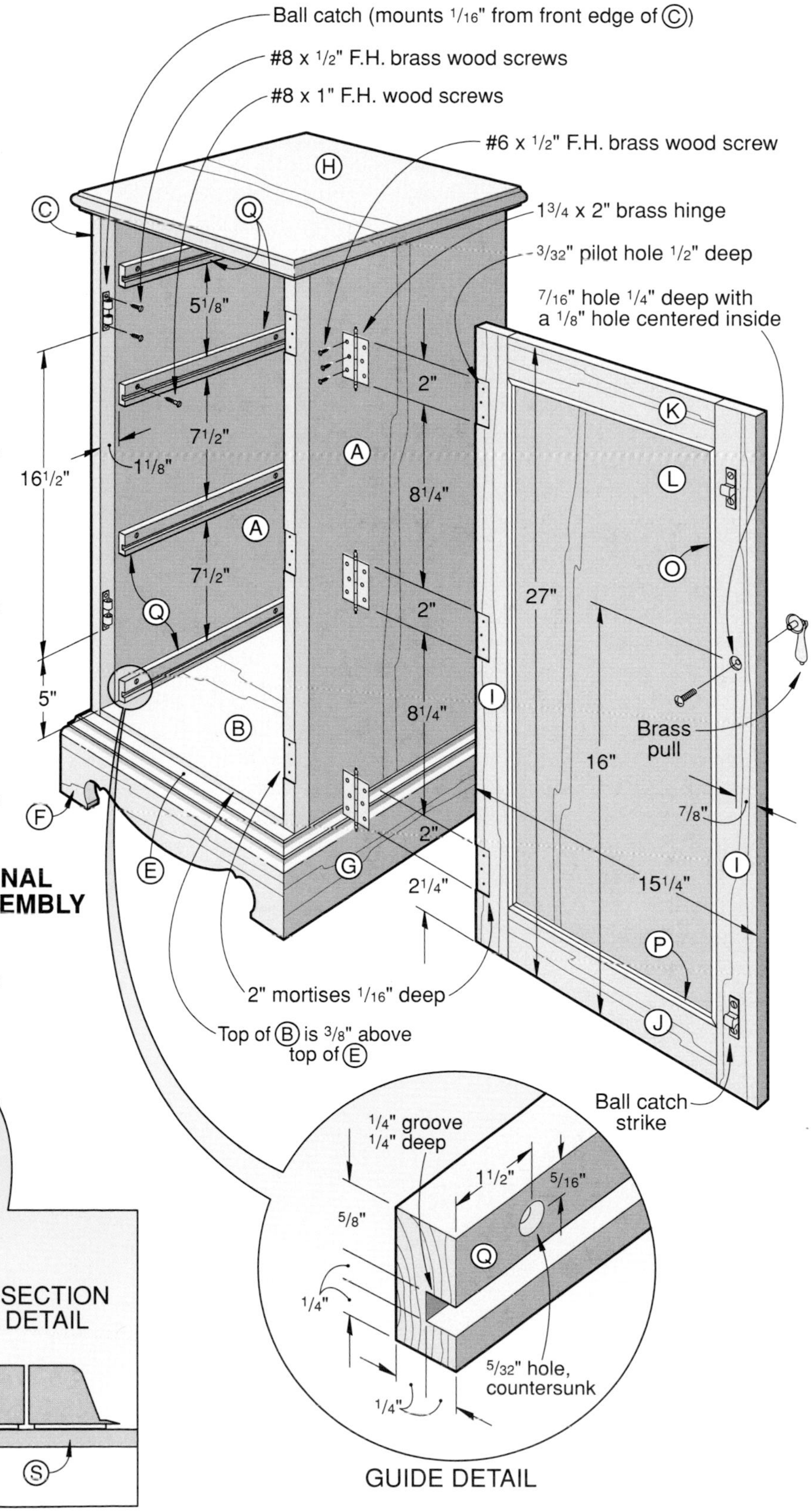

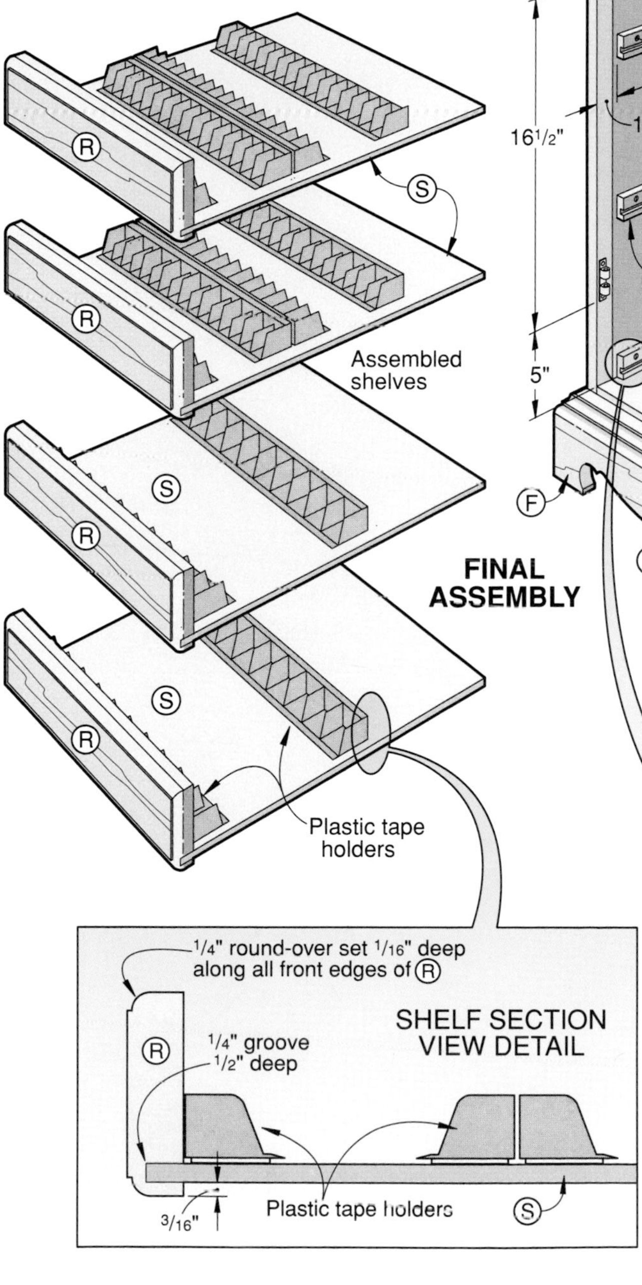

Assembled shelves

Plastic tape holders

FINAL ASSEMBLY

Ball catch (mounts $\frac{1}{16}$" from front edge of Ⓒ)

#8 x $\frac{1}{2}$" F.H. brass wood screws

#8 x 1" F.H. wood screws

#6 x $\frac{1}{2}$" F.H. brass wood screw

$1\frac{3}{4}$ x 2" brass hinge

$\frac{3}{32}$" pilot hole $\frac{1}{2}$" deep

$\frac{7}{16}$" hole $\frac{1}{4}$" deep with a $\frac{1}{8}$" hole centered inside

5$\frac{1}{8}$"

2"

$1\frac{1}{8}$"

$7\frac{1}{2}$"

$8\frac{1}{4}$"

$16\frac{1}{2}$"

$7\frac{1}{2}$"

2"

$8\frac{1}{4}$"

5"

2"

27"

16"

Brass pull

$\frac{7}{8}$"

2$\frac{1}{4}$"

15$\frac{1}{4}$"

2" mortises $\frac{1}{16}$" deep

Top of Ⓑ is $\frac{3}{8}$" above top of Ⓔ

Ball catch strike

$\frac{1}{4}$" round-over set $\frac{1}{16}$" deep along all front edges of Ⓡ

$\frac{1}{4}$" groove $\frac{1}{2}$" deep

SHELF SECTION VIEW DETAIL

Plastic tape holders

$\frac{3}{16}$"

$\frac{1}{4}$" groove $\frac{1}{4}$" deep

$\frac{5}{8}$"

$1\frac{1}{2}$"

$\frac{5}{16}$"

$\frac{1}{4}$"

$\frac{1}{4}$"

$\frac{5}{32}$" hole, countersunk

$\frac{1}{4}$"

GUIDE DETAIL

Show-Off Showcase

The clear choice for displaying your collectibles

Half the fun of collecting is showcasing your pieces for viewing. But, you don't want the display case to take away from the beauty of your collectibles. With that in mind, I designed this simple but sturdy oak-framed case. I kept the stiles and rails narrow to maximize the view, and the shelves adjust up or down to fit your needs.

James R. Downing
Design Editor

LET'S START WITH THE BACK, TOP, AND BOTTOM

1 Cut the cabinet back (A) to the size listed in the Bill of Materials from ¾" oak plywood. Cut the top and bottom (B) to size from ¾"-thick solid-oak stock.

2 Cut or rout ⅜" rabbets ¼" deep along all *front and back* edges of the plywood back panel (A) where shown on the Exploded View drawing and accompanying Rabbet and Groove Detail. Cut a pair of the same-sized rabbets across the ends of the top and bottom pieces (B).

3 Cut a ¼" groove ⅜" deep ¼" from the back edge on the inside face of the top and bottom pieces. Dry-fit the pieces (A, B) to check the fit.

4 For hanging the cabinet on the wall later, drill a pair of ³⁄₁₆" countersunk shank holes through the top of the back (A) where shown on the Exploded View drawing.

CONSTRUCT THE SIDE FRAMES AND DOOR NEXT

1 For the side frame stiles (C) and rails (D), cut four pieces of ¾"-thick stock to 1" wide by 38" long. Cut the oak door frame pieces (E, F) to size plus 1" in length each.

2 Using steps 1 and 2 of the 3-step drawing titled Forming the Door and Side Frame Parts for reference, rout and cut the strips for the side frame stiles and rails (C, D) to shape. Repeat the process using all three steps on the drawing to form the strips for the door stiles and rails (E, F).

3 Miter-cut the side frame stiles (C) and rails (D) to length.

4 Glue and clamp each set of side frames together, checking for square and flatness. (We found it helpful to use band clamps to hold the mitered ends tight until the glue dried.)

5 Build the Spline-Cutting Jig shown *below*. As shown in the drawing, raise the saw blade 1" above the saw table. To center the spline kerf in the corner edges of the frame, position the fence ⅝" from the center of the blade. (This takes into consideration the ¼" hardboard side of the jig.) As shown in the photo *below left*, place a corner of one side frame in the jig, start the saw, and cut a spline kerf centered in the corner of the frame. (We test-cut scrap stock the same thickness as the side frame first to verify the cut would be centered.) See the Side Frame drawing for reference. Repeat for each corner of each frame. *continued*

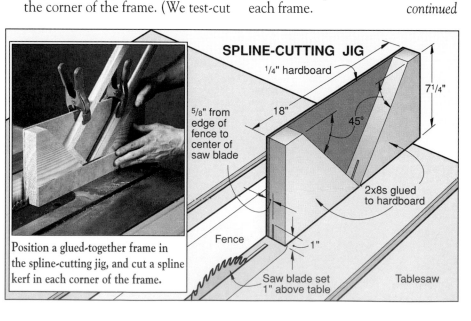

Position a glued-together frame in the spline-cutting jig, and cut a spline kerf in each corner of the frame.

SPLINE-CUTTING JIG

¼" hardboard

7¼"

⅝" from edge of fence to center of saw blade

18"

45°

2x8s glued to hardboard

Fence

1"

Saw blade set 1" above table

Tablesaw

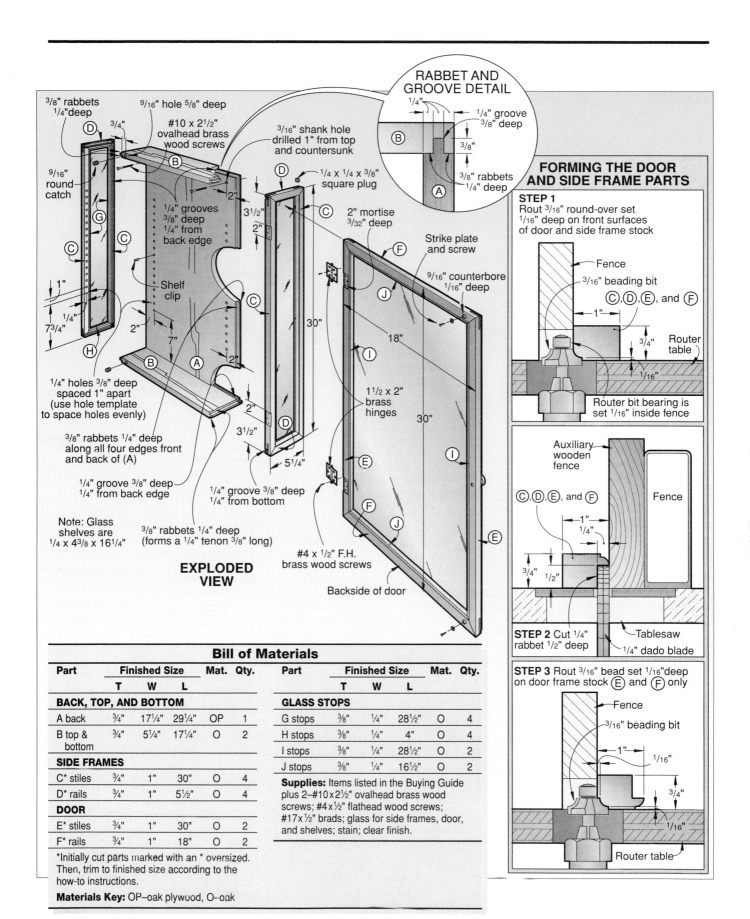

RABBET AND GROOVE DETAIL

1/4"

1/4" groove
3/8" deep

3/8"

3/8" rabbets
1/4" deep

FORMING THE DOOR AND SIDE FRAME PARTS

STEP 1
Rout 3/16" round-over set 1/16" deep on front surfaces of door and side frame stock

Fence

3/16" beading bit

C, D, E, and F

1"

3/4"

Router table

1/16"

Router bit bearing is set 1/16" inside fence

Auxiliary wooden fence

C, D, E, and F

Fence

1"

1/4"

3/4"

1/2"

STEP 2 Cut 1/4" rabbet 1/2" deep

Tablesaw

1/4" dado blade

STEP 3 Rout 3/16" bead set 1/16" deep on door frame stock E and F only

Fence

3/16" beading bit

1"

1/16"

3/4"

1/16"

Router table

3/8" rabbets 1/4" deep

3/4"

9/16" hole 5/8" deep

#10 x 2 1/2" ovalhead brass wood screws

B

9/16" round catch

G

C

C

C

1/4" grooves 3/8" deep 1/4" from back edge

Shelf clip

1"

1/4"

7 3/4"

H

2"

2"

7"

B

A

2"

1/4" holes 3/8" deep spaced 1" apart (use hole template to space holes evenly)

3/8" rabbets 1/4" deep along all four edges front and back of (A)

1/4" groove 3/8" deep 1/4" from back edge

Note: Glass shelves are 1/4 x 4 3/8 x 16 1/4"

3/8" rabbets 1/4" deep (forms a 1/4" tenon 3/8" long)

EXPLODED VIEW

3/16" shank hole drilled 1" from top and countersunk

D

1/4 x 1/4 x 3/8" square plug

C

3 1/2"

2"

2" mortise 3/32" deep

30"

C

2"

D

3 1/2"

5 1/4"

E

#4 x 1/2" F.H. brass wood screws

F

Strike plate and screw

9/16" counterbore 1/16" deep

J

18"

I

1 1/2 x 2" brass hinges

30"

I

F

J

E

Backside of door

Bill of Materials

Part	Finished Size			Mat.	Qty.
	T	W	L		
BACK, TOP, AND BOTTOM					
A back	3/4"	17 1/4"	29 1/4"	OP	1
B top & bottom	3/4"	5 1/4"	17 1/4"	O	2
SIDE FRAMES					
C* stiles	3/4"	1"	30"	O	4
D* rails	3/4"	1"	5 1/2"	O	4
DOOR					
E* stiles	3/4"	1"	30"	O	2
F* rails	3/4"	1"	18"	O	2

Part	Finished Size			Mat.	Qty.
	T	W	L		
GLASS STOPS					
G stops	3/8"	1/4"	28 1/2"	O	4
H stops	3/8"	1/4"	4"	O	4
I stops	3/8"	1/4"	28 1/2"	O	2
J stops	3/8"	1/4"	16 1/2"	O	2

Supplies: Items listed in the Buying Guide plus 2–#10 x 2 1/2" ovalhead brass wood screws; #4 x 1/2" flathead wood screws; #17 x 1/2" brads; glass for side frames, door, and shelves; stain; clear finish.

*Initially cut parts marked with an * oversized. Then, trim to finished size according to the how-to instructions.

Materials Key: OP–oak plywood, O–oak

Show-Off Showcase

continued

6 Cut eight pieces of ⅛" oak to 1½" wide by 2½" long. Glue one of the splines in each kerfed corner of both frames. Later, trim and sand the splines flush with the outside edges of the frames.

7 To reduce the thickness of the front stile (C), rip ¼" off the front edge of each side frame. Left untrimmed, the assembled cabinet with the door attached would look front heavy.

8 Cut a ¼" groove ⅜" deep ¼" from the back and bottom edges (not the front edge) of each assembled side frame.

9 Miter-cut the door-frame pieces (E, F) to length, glue them together, and kerf and spline the corners just like you did for the pair of side frames. See the Door drawing for reference.

10 Mark the centerpoint, and drill an ¹¹⁄₆₄" hole in the door for adding the brass knob later.

LET THE ASSEMBLY BEGIN

1 Have single-strength glass cut to size for the side panels and door. At the same time, have as many ¼ × 4⅜ × 16¼" glass shelves cut to size as you'll need.

2 Mark the locations and form the two ³⁄₃₂"-deep hinge mortises in the right-hand side frame stile (C) where dimensioned on the Exploded View drawing. With the top and bottom edges flush, lightly clamp the door frame to the right-hand stile (C). Mark lines to transfer the hinge locations from the edge of the side frame stile across the back surface of the door stile (E). Remove the door, and rout and chisel the mortises where marked. Using the hinges as guides, drill pilot holes in the pieces for mounting the hinges later. (See the Buying Guide for our source of hardware.)

3 To mount the round magnetic catches, drill a pair of ⁹⁄₁₆" holes ⅝" deep in the top and bottom

pieces (B) where shown on the Exploded View drawing. Drill a test hole in scrap stock first to verify the hole size and depth for your particular catches. Only ¹⁄₁₆" of the end of the catch should protrude past the front edge of the top and bottom pieces. Install the catches.

4 Dry-clamp (no glue) the top and bottom pieces (B) and side frames (C, D) to the back panel (A). The front edges and top and bottom ends should be flush; trim if necessary. Glue and clamp the assembly. Check for square and remove any excess glue.

5 Note that you have a ¼ × ⅜" notch at each corner of the cabinet where shown on the Exploded View drawing. Cut four filler blocks to fit the notches, and glue them in place.

6 Construct a template like that shown *above right*. Using a stop for consistent depth and the template for even spacing, drill ¼" holes ⅜" deep in the back panel (A) and front side frame stiles (C) where shown on the Exploded View drawing.

7 Using brass screws, screw the hinges to the door and then to the cabinet. Put a screw through each strike plate, and hold it against the magnetic catch. Close the door, and squeeze the door and screw against the cabinet to indent the strike-plate locations to the door rails. Drill mounting holes and add the strike plates to the door.

8 From ⅜" stock (we planed thicker stock to this thickness) rip enough ¼"-wide strips for glass stops (G, H, I, J). Miter-cut the stops to length.

9 Drill pilot holes through the stops for nailing them in place later. To do this, snip the head off a #17 × ½" brad. Chuck the headless brad into your portable drill. Use the drill and "pilot bit" to drill the pilot holes through the stops.

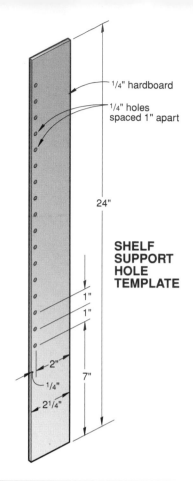

¼" hardboard

¼" holes spaced 1" apart

24"

SHELF SUPPORT HOLE TEMPLATE

1"

1"

2"

¼"

2¼"

7"

FINAL CLEANUP AND ASSEMBLY

1 Remove the hinges and strike plates. Finish-sand the cabinet, the door, and the glass stops. Mask the magnetic catches, then stain the pieces and add a clear finish.

2 Using #17 × ½" brads, secure the glass in the side frames with the glass stops. Repeat the process to secure the glass panel in the door.

3 Attach the brass knob. Reattach the hinges to secure the door to the cabinet. Then, remove the hinge pin from each hinge to separate the door from the cabinet (it's easier to hang the cabinet without the door). With a helper, level and hang the cabinet.

4 Position the shelf clips and glass shelves. Reattach the strike plates, and add the hinge pins to reconnect the door to the cabinet.

BUYING GUIDE

• **Hardware kit.** One pair of 1½ × 2" ball-tipped solid-brass hinges (VH2014), 1"-diam. solid brass knob (SBH14), 2–⁹⁄₁₆"-diam. magnetic catches and strike plates (55F25), shelf clips (HAF284). For current prices contact Constantine's, 2050 Eastchester Road, Bronx, NY 10461, or call 1-800-223-8087 to order.

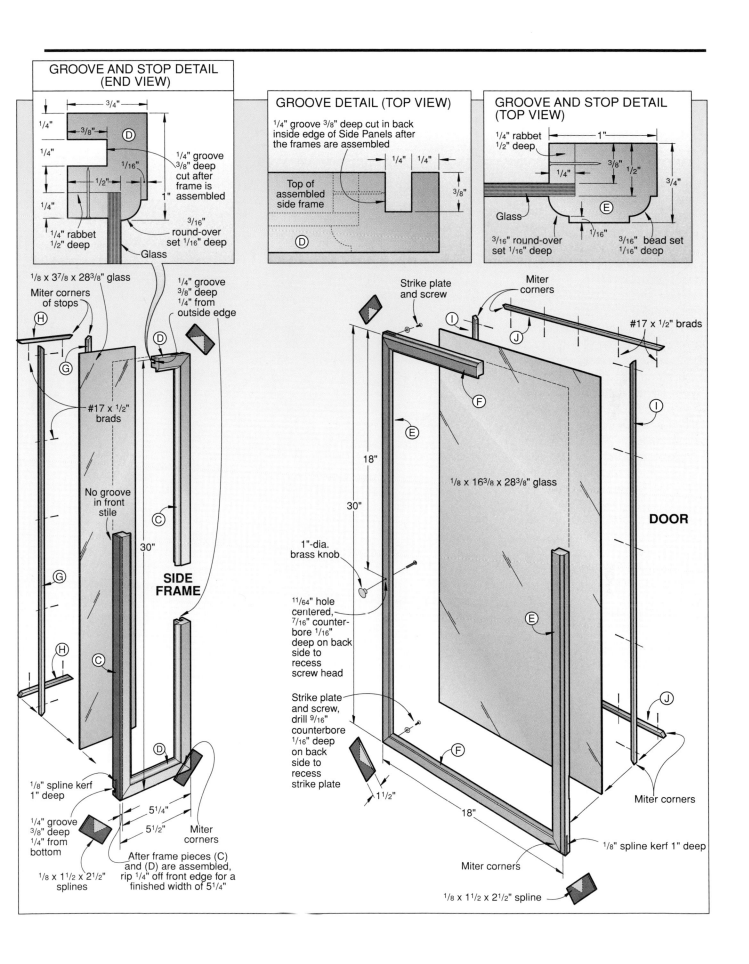

GROOVE AND STOP DETAIL (END VIEW)

GROOVE DETAIL (TOP VIEW)

GROOVE AND STOP DETAIL (TOP VIEW)

Corner Classic

A traditionally-styled oak cabinet with plenty of personality

Family rooms have them, and dining rooms too—drab, life-less corners that serve little or no purpose. But with the cabinet featured here, you can say good-bye to dull decorating and hello to one classy furniture piece.

FORMING THE CARCASE

1 Rip and crosscut the back panels (A, B) to size from ¾" oak plywood. Because of the joinery used, panel B is slightly narrower than panel A. For ease in machining and assembling the two different-sized panels later, mark an A on the wider panel and a B on the other. (We stuck a piece of masking tape about 1" long on each panel and wrote the letter designation on the tape. Doing this prevents you from having to sand a penciled-on letter off the thin-veneered surface later.)

2 From ¾"-thick oak, cut the sides (C) to size *plus ¼" in width*. Mark an X on the outside face of each side panel.

3 Using the Carcase Assembly and Side Panel drawings *(opposite and on page 66)* for reference, lay out the dado locations on the front of each back panel (A, B) and the inside of the sides (C).

4 With a ¾" straight bit, rout ¼"-deep dadoes in each back panel. Clamp the side panels (C) edge to edge, with the good face down, and rout the dadoes across them. Cut a ¾" rabbet ¼" deep across the top end of the back and side panels.

5 Cut a ⅜" groove ⅜" deep ⅜" in from the back edge of back panel A where shown on the Carcase Assembly drawing. Cut a ⅜" rabbet ⅜" deep along the left-hand edge of back panel A and along *both* edges of back panel B where shown on the same drawing.

continued

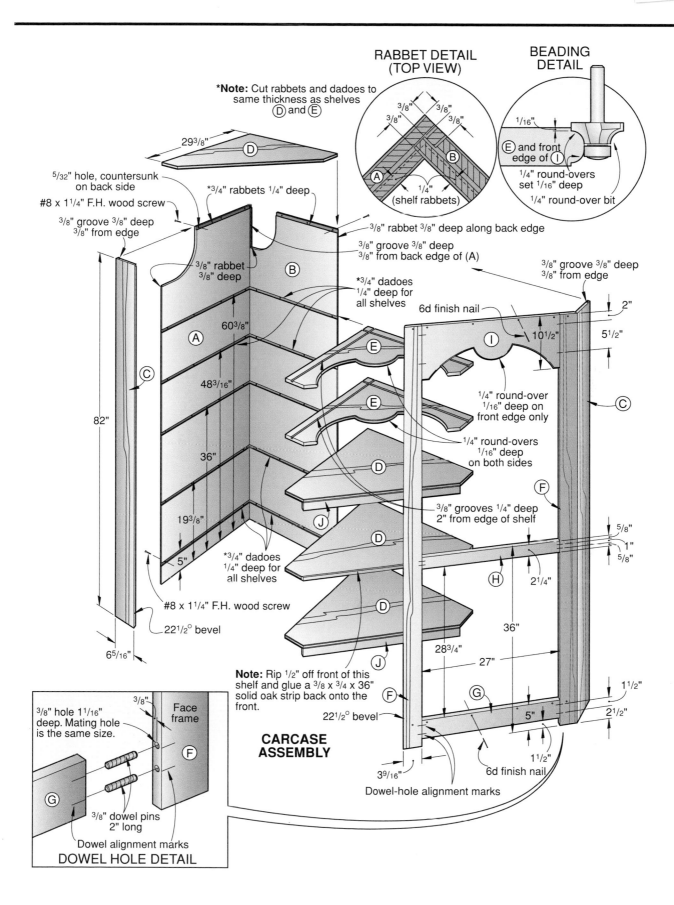

RABBET DETAIL
(TOP VIEW)

3/8" 3/8"
3/8" 3/8"

Ⓐ Ⓑ

1/4"
(shelf rabbets)

BEADING
DETAIL

1/16"

Ⓔ and front
edge of Ⓘ

1/4" round-overs
set 1/16" deep

1/4" round-over bit

*Note: Cut rabbets and dadoes to
same thickness as shelves
Ⓓ and Ⓔ

29³/₈" Ⓓ

5/32" hole, countersunk
on back side

#8 x 1¹/₄" F.H. wood screw

3/8" groove 3/8" deep
3/8" from edge

*3/4" rabbets 1/4" deep

3/8" rabbet
3/8" deep

Ⓑ

Ⓐ

3/8" rabbet 3/8" deep along back edge

3/8" groove 3/8" deep
3/8" from back edge of (A)

3/8" groove 3/8" deep
3/8" from edge

2"

5¹/₂"

60³/₈"

48³/₁₆"

Ⓒ

*3/4" dadoes
1/4" deep for
all shelves

6d finish nail

Ⓘ

10¹/₂"

Ⓔ

Ⓔ

1/4" round-over
1/16" deep on
front edge only

1/4" round-overs
1/16" deep
on both sides

Ⓒ

82"

36"

Ⓓ

Ⓕ

3/8" grooves 1/4" deep
2" from edge of shelf

19³/₈"

*3/4" dadoes
1/4" deep for
all shelves

Ⓙ

5/8"
1"
5/8"

Ⓓ

5"

Ⓗ

2¹/₄"

#8 x 1¹/₄" F.H. wood screw

22¹/₂° bevel

Ⓓ

36"

Ⓙ

28³/₄"

6⁵/₁₆"

Note: Rip ¹/₂" off front of this
shelf and glue a ³/₈ x ³/₄ x 36"
solid oak strip back onto the
front.

27"

1¹/₂"

Ⓕ

Ⓖ

2¹/₂"

CARCASE
ASSEMBLY

22¹/₂° bevel

5"

3⁹/₁₆"

6d finish nail

1¹/₂"

Dowel-hole alignment marks

3/8" hole 1¹/₁₆"
deep. Mating hole
is the same size.

3/8"

Face
frame

Ⓕ

Ⓖ

3/8" dowel pins
2" long

Dowel alignment marks

DOWEL HOLE DETAIL

Corner Classic

continued

6 Cut a ⅜" groove ⅜" deep ⅜" in from the back inside face of each side panel (C).

LET'S CUT THE SHELVES TO SHAPE

1 Using the Shelf Parts View drawing on *page 68* and the Cutting Diagram on *page 69* for reference, mark the outline for one shelf (D) on ¾" oak plywood.

2 Carefully cut the marked shelf to shape, and use it as a template to mark the outlines for the remaining three plywood shelves.

3 Rip ½" off the front edge of *one* plywood shelf, and glue a ⅜×¾×36" strip of solid oak onto the front of the shelf to hide the plies. Later, trim the ends of the strip flush with those of the shelf. (See the Carcase Assembly drawing on *page 65* for reference.)

4 Edge-join enough solid-oak stock for the two remaining shelves (E). Again, using the first shelf as a template, transfer the outline to the solid-oak shelves, and cut them to shape.

5 Enlarge the shelf half-pattern on *page 71* and transfer it to poster-board (for instructions on how to enlarge gridded patterns see *page 160*). Mark a centerline down the center of each solid-oak shelf. Cut the poster board to shape, and use it as a template to mark both halves of the front edge of the solid-oak shelves (E) as shown in the photo *at right*.

6 Cut the shelf fronts to shape and sand the cut edges to remove the saw marks. (For a smooth cut, we used a jigsaw fitted with a finish-cutting blade.)

7 To stand plates upright on the solid-oak shelves (E) later, cut ⅜" grooves ¼" deep 2" in from the back edges.

continued

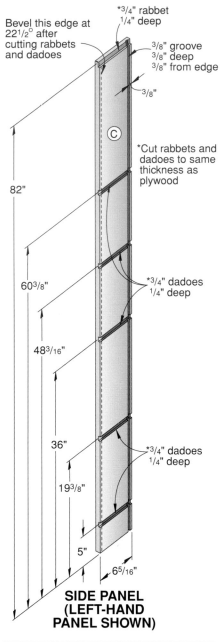

SIDE PANEL (LEFT-HAND PANEL SHOWN)

Labels on drawing:
- Bevel this edge at 22½° after cutting rabbets and dadoes
- *¾" rabbet ¼" deep
- ⅜" groove ⅜" deep ⅜" from edge
- ⅜"
- Ⓒ
- *Cut rabbets and dadoes to same thickness as plywood
- 82"
- 60³⁄₈"
- 48³⁄₁₆"
- 36"
- 19³⁄₈"
- 5"
- 6⁵⁄₁₆"
- *¾" dadoes ¼" deep
- *¾" dadoes ¼" deep

Transfer the full-sized shelf half pattern to poster board and use it to mark the front edge of the top two shelves.

Bill of Materials

Part	Finished Size*			Mat.	Qty.
	T	W	L		
CARCASE ASSEMBLY					
A back	¾"	30"	82"	OP	1
B back	¾"	29⁵⁄₈"	82"	OP	1
C* sides	¾"	6⁵⁄₁₆"	82"	O	2
D shelves	¾"	24⁷⁄₁₆"	41½"	OP	4
E shelves	¾"	24⁷⁄₁₆"	41½"	EO	2
FACE FRAME					
F stiles	¾"	3⁹⁄₁₆"	82"	O	2
G bottom rail	¾"	3½"	27"	O	1
H center rail	¾"	2¼"	27"	O	1
I top rail	¾"	10½"	27"	EO	1
J cleats	¾"	2"	34⅛"	O	2
BASE MOLDING					
K* inner front	¾"	4¾"	34¾"	O	1
L* inner sides	¾"	4¾"	6⅝"	O	2
M* outer front	¾"	4"	35⅜"	O	1
N* outer sides	¾"	4"	6¹⁵⁄₁₆"	O	2
UPPER MOLDING					
O* inner front	¾"	4⅛"	34¾"	O	1
P* inner sides	¾"	4⅛"	6⅝"	O	2
Q* outer front	crown mold	36⁷⁄₁₆"		O	1
R* outer sides	crown mold	7½"		O	2
WAIST MOLDING					
S* inner front	¾"	1¼"	35⅛"	O	1
T* inner sides	¾"	1¼"	6⅞"	O	2
U* front cove mold	¾"	¾"	34¾"	O	1
V* side cove mold	¾"	¾"	6⅝"	O	2
DOORS					
W stiles	¾"	3"	28¾"	O	4
X rails	¾"	3"	7½"	O	2
Y rails	¾"	4"	7½"	O	2
Z panels	¼"	7⁷⁄₁₆"	21¹¹⁄₁₆"	EO	2
AA* stops	¼"	¼"	21¾"	O	4
BB* stops	¼"	¼"	7½"	O	4
CC* stops	¼"	⅜"	21¾"	O	4
DD* stops	¼"	⅜"	7½"	O	4

*Initially cut parts marked with an * oversized. Then, trim to finished size according to the how-to instructions.

Materials Key: OP–oak plywood, O–oak, EO–edge-joined oak.

Supplies: ⅜" dowel pins 2" long, #18×⅝" brads, #8×1¼" flathead wood screws, #4d and #6d finish nails, wood putty, stain, clear finish.

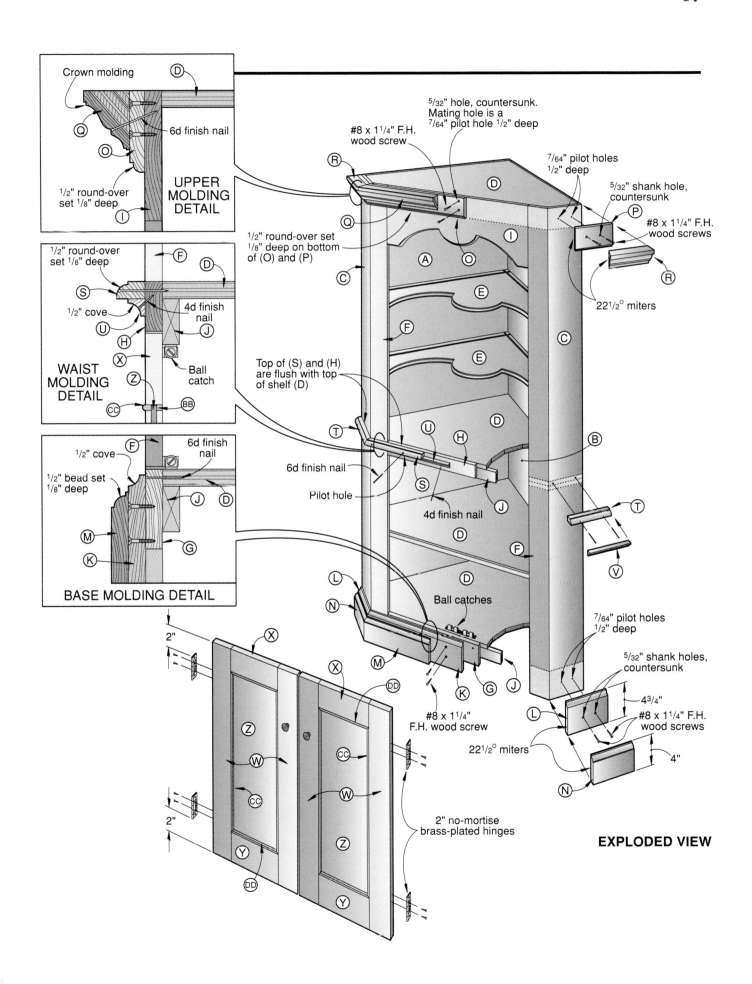

UPPER MOLDING DETAIL

Crown molding

6d finish nail

Q

O

I

1/2" round-over set 1/8" deep

WAIST MOLDING DETAIL

1/2" round-over set 1/8" deep

1/2" cove

4d finish nail

Ball catch

F

D

S

U

H

X

J

Z

CC

BB

BASE MOLDING DETAIL

1/2" cove

6d finish nail

1/2" bead set 1/8" deep

F

D

J

M

G

K

5/32" hole, countersunk. Mating hole is a 7/64" pilot hole 1/2" deep

#8 x 1 1/4" F.H. wood screw

7/64" pilot holes 1/2" deep

5/32" shank hole, countersunk

#8 x 1 1/4" F.H. wood screws

22 1/2° miters

1/2" round-over set 1/8" deep on bottom of (O) and (P)

Top of (S) and (H) are flush with top of shelf (D)

6d finish nail

Pilot hole

4d finish nail

Ball catches

#8 x 1 1/4" F.H. wood screw

2" no-mortise brass-plated hinges

7/64" pilot holes 1/2" deep

5/32" shank holes, countersunk

4 3/4"

#8 x 1 1/4" F.H. wood screws

4"

22 1/2° miters

2"

2"

EXPLODED VIEW

Corner Classic

continued

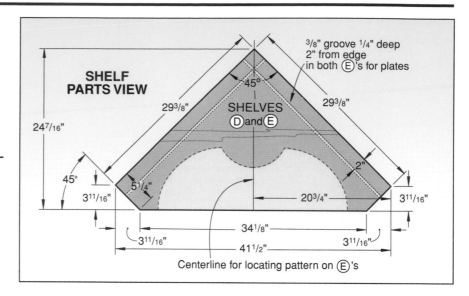

SHELF PARTS VIEW

SHELVES Ⓓ and Ⓔ

45°

3/8" groove 1/4" deep 2" from edge in both Ⓔ's for plates

29 3/8"   29 3/8"

24 7/16"

45°

5 1/4"

3 11/16"   3 11/16"

2"

20 3/4"   3 11/16"

34 1/8"

3 11/16"   41 1/2"   3 11/16"

Centerline for locating pattern on Ⓔ's

8 Using the Beading Detail accompanying the Carcase Assembly drawing on *page 65* for reference, rout the curved front edges of the solid-oak shelves (E). Sand the routed edges.

ASSEMBLE THE BASIC CARCASE

1 Dry-clamp (no glue) the back panels (A, B), the sides (C), and shelves (D, E) to check the fit of all the pieces. Since the front edge of each side piece (C) is not yet beveled, just check the alignment of the dadoes with those machined in the back panels.

2 Glue and clamp the back panels (A, B) together. While the glue is still wet, install the shelves (D, E). Working from the back side, drill countersunk screw holes through the back panels and into the shelves. Using #8×1 1/4" flathead wood screws pull the shelves tight into the rabbets and dadoes. (To make excess glue removal easier, we applied masking tape next to the glue joints, and removed the tape [and glue stuck to it] about 15 minutes after assembly. Do not leave the tape on overnight; it's harder to remove and leaves a sticky residue.)

NOW, ADD THE FACE FRAME FOR A FINISHED FRONT

1 Cut the face frame stiles (F), bottom rail (G), center rail (H), and top rail (I) to the sizes listed in the Bill of Materials.

2 Dry-clamp the parts together, and lay out the dowel-hole alignment marks across each joint where dimensioned on the Carcase Assembly drawing.

3 Using the same method used to transfer the shape to the front of the shelves (E), enlarge and transfer the top rail half-pattern from the pattern on *page 71* to the top rail (I).

4 Remove the clamps, and cut the top rail to shape. Then, rout a bead like that used on the shelves (E) along the *front edge only* of the top rail where shown on the Carcase Assembly drawing on *page 65*.

5 Use a doweling jig to drill 3/8" holes 1 1/16" deep in the stiles and rails where previously marked. See the Dowel Hole Detail accompanying the Carcase Assembly drawing for reference.

6 Glue and clamp the face frame together, checking for square. Later, remove the clamps, scrape off the excess glue, and sand the front and back of the face frame assembly.

7 Tilt your tablesaw blade 22.5° from vertical. Add an auxiliary wooden fence to your rip fence, raise the blade into the wooden auxiliary fence where shown on the Bevel-Ripping the Face Frame drawing *above right*. Next, bevel-rip the outside edges of the face frame. You'll find it handy to have a helper at the outfeed end of your tablesaw to support the face frame when completing the cut. Dry-clamp the face frame to the carcase assembly and verify the necessary width for the sides (C). (As mentioned earlier,

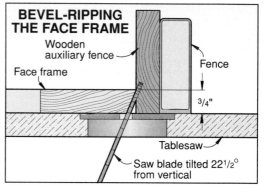

BEVEL-RIPPING THE FACE FRAME

Wooden auxiliary fence

Face frame

Fence

3/4"

Tablesaw

Saw blade tilted 22 1/2° from vertical

we had you cut the side panels to size *plus 1/4" in width*. The extra width comes in handy now when bevel-ripping the side panels to width for a perfect fit.)

8 Remove the auxiliary fence, and with the blade tilted *away* from the rip fence, bevel-rip the *front edge* of each side panel (C) to fit against the cabinet and face frame. Glue and clamp the side panels to the carcase.

JOIN THE FACE FRAME TO THE CARCASE

1 Cut the cleats (J) to size. Glue them to the bottom side of the two shelves where shown on the Carcase Assembly drawing.

2 Glue and clamp the face frame to the cabinet assembly. The cleats (J) are handy to clamp to in order to pull the face frame tight against the

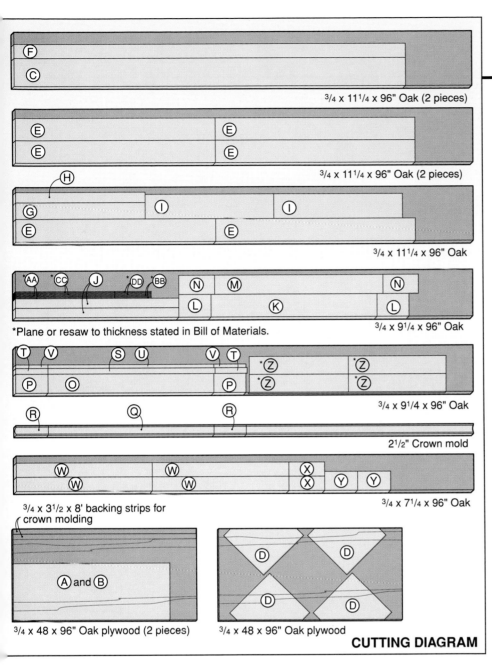

F
C

¾ x 11¼ x 96" Oak (2 pieces)

E   E
E   E

¾ x 11¼ x 96" Oak (2 pieces)

H
G   I   I
E   E

¾ x 11¼ x 96" Oak

*AA *CC J *DD *BB   N   M   N
L   K   L

*Plane or resaw to thickness stated in Bill of Materials.

¾ x 9¼ x 96" Oak

T V   S U   V T
P O   P   *Z   *Z
R   Q   R   *Z   *Z

¾ x 9¼ x 96" Oak

R   Q   R

2½" Crown mold

W   W   X
W   W   X Y Y

¾ x 3½ x 8' backing strips for crown molding

¾ x 7¼ x 96" Oak

A and B

¾ x 48 x 96" Oak plywood (2 pieces)

D   D
D
D   D

¾ x 48 x 96" Oak plywood

CUTTING DIAGRAM

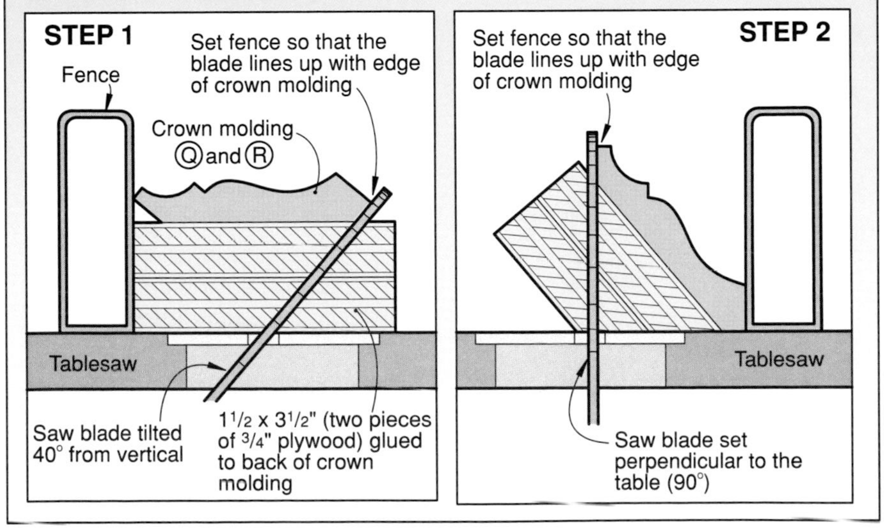

RIPPING THE CROWN MOLDING

STEP 1

Fence

Set fence so that the blade lines up with edge of crown molding

Crown molding Q and R

Tablesaw

Saw blade tilted 40° from vertical

1½ x 3½" (two pieces of ¾" plywood) glued to back of crown molding

STEP 2

Set fence so that the blade lines up with edge of crown molding

Tablesaw

Saw blade set perpendicular to the table (90°)

front of the cabinet. If necessary for a gap-free joint, use 6d finish nails to further secure the stiles (F) to the sides (C).

THE WAIST, AND UPPER MOLDINGS COME NEXT

Note: *The lengths of parts K through V are all based on the dimensions of the cabinet shown here. Since it's quite easy to get off a fraction of an inch when building your cabinet, measure the length needed for your trim parts before cutting to length.*

Because of limited surfaces to clamp to and to allow for wood movement of the solid-wood sides (C), we nailed the trim pieces in place, using nails sparingly.

1 Cut a 4¾"-wide piece of ¾" oak to 52" for the inner base molding pieces (K, L). (We found it easier when making the trim pieces to cut one lineal length, rout the edge as necessary, measure the cabinet to determine the length of the piece needed, and then cut the trim from the lineal stock.)

2 Rout a ½" cove along the top front edge of this 52"-long piece where shown on the Base Molding Detail accompanying the Exploded View drawing on *page 67.* Miter-cut the pieces to length, drill countersunk screw holes, and screw (no glue) the pieces to the cabinet.

3 Cut a ¾ x 4 x 54" piece of stock for the outer base molding pieces (M, N). Rout a ½" bead set ⅛" deep along the top front edge. Cut the pieces to length, and glue and clamp them to the cabinet.

4 Cut a ¾" piece of oak stock for the inner upper molding pieces (O, P). Rout a ½" round-over set ⅛" deep along one edge of the stock where shown on the Upper Molding Detail accompanying the Exploded View drawing. Cut the pieces to length, drill countersunk screw holes and screw (no glue) the pieces to the cabinet.

continued

Corner Classic

continued

5 Buy an 8' length of oak crown molding. Then, rip two pieces of ¾" plywood to 3½" wide by 8' long. Glue the plywood and the crown molding face-to-face in the configuration shown on the drawing titled Ripping the Crown Molding on *page 69*.

6 Follow the two steps in the Ripping the Crown Molding drawing to bevel-rip the edges of the plywood flush with the edges of the crown molding.

7 Cut the crown molding pieces (Q, R) to length, and glue and nail them to the top of the cabinet. See the Upper Molding detail accompanying the Exploded View drawing for reference.

8 Cut a ¾ x 1¼ x 52" strip for the waist molding pieces (S, T). Rout a ½" round-over set ⅛" deep along the outside edge of the strip where shown on the Waist Molding Detail accompanying the Exploded View drawing. Then, cut the pieces (S, T) to length, and nail them to the cabinet.

9 Rout the ½" cove molding for parts U, V (you can also purchase preformed ½" oak cove molding). Cut the pieces to length, and nail them to the cabinet.

A PAIR OF DOORS ENCLOSE THE BOTTOM

Note: *The dimensions for the door stiles and rails will provide you with a pair of doors that fit in the opening leaving no gap. After building the doors, trim the edges as necessary to accommodate the hinges and create a ¹⁄₁₆" gap top and bottom. We find it easier to build the doors to fill the opening, and trim as necessary.*

1 Cut the door stiles (W) and rails (X, Y) to size. Dry-clamp each door together, and transfer the dowel-hole alignment marks across each joint where shown on the Door assembly drawing.

2 Drill the dowel holes. Glue and dowel each door, checking for square and flatness.

3 Edge-join ¼" oak (we planed thicker stock) to form the two ¼"-thick door panels (Z). Later, trim the panels to finished size. Finish-sand the panels, and stain them. (Staining at this point ensures that you won't have any unstained strips along the outside edges later when the panels have had a chance to contract in the frames.)

4 Cut two ¼ x ¼ x 7' lineal lengths for the back panel stops (AA, BB) and two ¼ x ⅜ x 6' strips for the front panel stops (CC, DD). Rout the front edges of the front panel stop strips where shown on the Section View Detail *above*.

5 Miter-cut the stops to length, stain them, and nail them in place to secure a panel in each door.

6 Mark the centerpoint and drill a hole through each door for mounting the knobs later. (See the Door drawing *above* for positioning particulars).

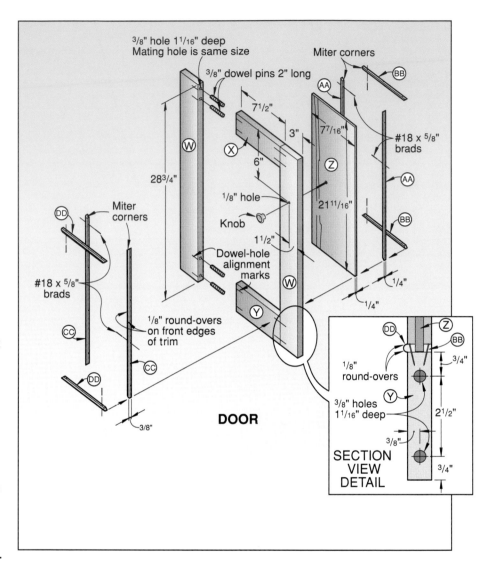

DOOR

3/8" hole 1 1/16" deep
Mating hole is same size

3/8" dowel pins 2" long

Miter corners

7 1/2"

3"

7 7/16"

#18 x 5/8" brads

28 3/4"

6"

1/8" hole

Knob

21 11/16"

1/4"

1/4"

1 1/2"

Dowel-hole alignment marks

Miter corners

#18 x 5/8" brads

1/8" round-overs on front edges of trim

3/8"

SECTION VIEW DETAIL

1/8" round-overs

3/8" holes 1 1/16" deep

3/4"

2 1/2"

3/8"

3/4"

7 Attach no-mortise hinges (see the Buying Guide for our source of hardware) to the cabinet and doors. Trim as necessary to fit the doors into the opening with $\frac{1}{16}$" gaps top and bottom. Attach the ball catches and strikes to the doors and cabinet where shown on the Waste Trim and Base Molding details accompanying the Exploded View drawing.

FINAL CLEANUP, THEN IT'S OFF TO THE CORNER

1 Remove all the hardware from the cabinet and doors. Set all the nails, and putty the holes. Finish-sand the cabinet and each door.

2 Wipe on the stain. (For a traditional look, we used Bartley's walnut gel stain. Call 1-800-787-2800 for the dealer nearest you.) Use a clean cotton cloth to apply the stain to one section at a time (don't try to do the whole thing at once). Wipe off the excess immediately, being careful to wipe off with the grain. If you prefer a darker color, stain again after six hours. Buff lightly for an even color. Apply a clear finish (we brushed on three coats of Minwax Fast-Drying satin polyurethane).

3 Reattach the hardware and hang the doors. For lighting inside the upper cabinet (which is helpful in a dimly lit corner), add a cabinet light as sourced *below*.

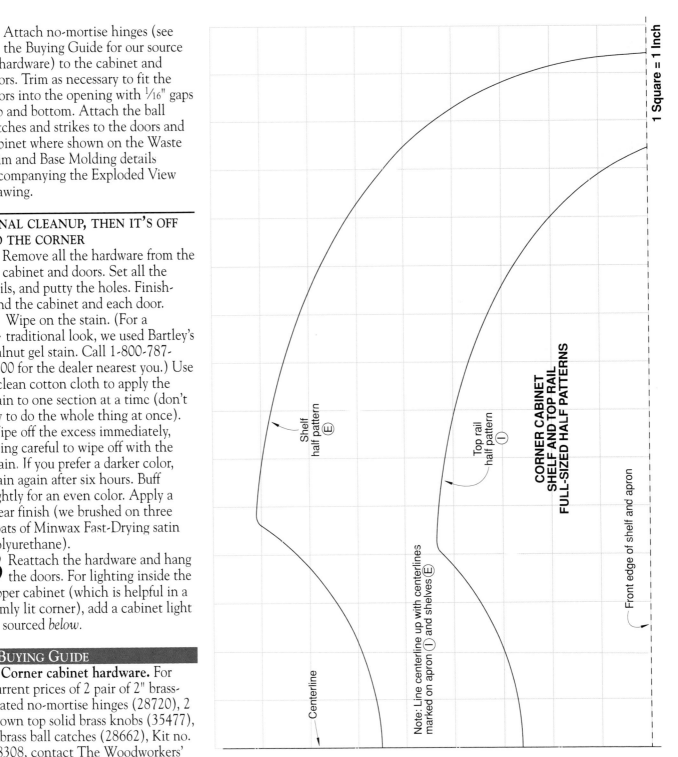

1 Square = 1 Inch

Shelf half pattern E

Top rail half pattern I

CORNER CABINET SHELF AND TOP RAIL FULL-SIZED HALF PATTERNS

Front edge of shelf and apron

Centerline

Note: Line centerline up with centerlines marked on apron I and shelves E

Young-at-Heart Projects

As soon as youngsters see these hierloom quality toys a smile is guaranteed. And their smiles are often contagious, especially when you know you've made a gift that will last for generations of kids to come. From the Scrollsawed Safari Puzzle to The Fun-Time Racer, all of these projects are designed to dazzle the young-at-heart.

A Reel Deal for Kids

Land a whopping "Gee, thanks!" with this tackle

Here's all the gear a young angler needs to lure the big ones out of the playroom fishin' hole. Building it? Well, that's not much more work than tossing a hook into a cool, clear lake on a sunny Sunday afternoon.

Note: *Dowel diameters may vary from the nominal size, so select drill bits to match the diameter of your ⅛", ¼", and ½" dowels. Use those bits wherever the instructions call for ⅛", ¼", or ½" holes.*

To build the fishing pole, you'll have to drill into the side of a few dowels (crosswise) and into the end of some (lengthwise). To make those tasks easier and safer, start by building the two jigs shown on *page 75* to use with your drill press.

For the rod, cut parts A, B, and C to length. (We cut the short pieces of dowel by hand, holding them in the V-block and cutting them with a fine-tooth backsaw.)

Drill the holes in parts A and B where shown. To drill lengthwise, position the dowel in the V-groove of the drilling fixture. Secure the

dowel with a clamp. Now, locate the center of the dowel, and place it under the bit. Clamp the base of the jig to the drill-press table, and drill the hole.

For the crosswise hole, clamp the V-block to your drill-press table with the center of the groove under the bit. Lay the dowel in the jig, and drill. After drilling part B, cut the end nearest the cross hole at a 45° angle. Glue parts A, B, and C together.

Drill four equally spaced holes in a pair of 2" wheels where shown by the Reel Side Detail. Cut parts D, E, and F to length. Glue parts D into one wheel, extending out from the back of the wheel (the flat side). Slide one part D through the hole in part B, install the other wheel, and center the reel on the rod. Glue the assembly in place in part B.

Refer to the Crank Hub and Shaft Retainer drawing, and follow the three steps shown to make parts G and H. Cut parts I, J, and K to size. Drill parts E, I, and K where shown, and assemble the reel. Enlarge the wheel center holes to allow part E to rotate freely. Slide E into position, and glue parts G and H to it, leaving E free to turn.

Glue parts F and I together, then slide a spandrel ball (a 1" ball with a ⅜" hole through it) over I. Secure the ball with part J. Grip a ¾"-

continued

A Reel Deal for Kids

continued

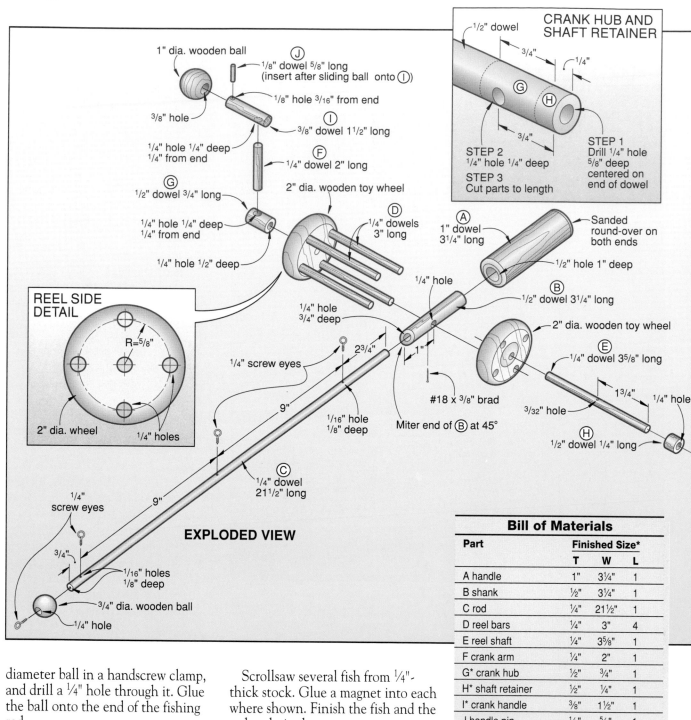

CRANK HUB AND SHAFT RETAINER

½" dowel
¾"
¼"
Ⓖ
Ⓗ
¾"

STEP 2
¼" hole ¼" deep
STEP 3
Cut parts to length

STEP 1
Drill ¼" hole
⅝" deep
centered on
end of dowel

1" dia. wooden ball

Ⓙ
⅛" dowel ⅝" long
(insert after sliding ball onto Ⓘ)

⅛" hole ³⁄₁₆" from end

⅜" hole

Ⓘ
³⁄₈" dowel 1½" long

¼" hole ¼" deep
¼" from end

Ⓕ
¼" dowel 2" long

Ⓖ
½" dowel ¾" long

2" dia. wooden toy wheel

¼" hole ¼" deep
¼" from end

Ⓓ
¼" dowels
3" long

Ⓐ
1" dowel
3¼" long

Sanded
round-over on
both ends

¼" hole ½" deep

½" hole 1" deep

Ⓑ
½" dowel 3¼" long

¼" hole
¾" deep

¼" hole

2" dia. wooden toy wheel

Ⓔ
¼" dowel 3⅝" long

REEL SIDE DETAIL

R=⅝"

2" dia. wheel   ¼" holes

¼" screw eyes

2¾"

9"

¹⁄₁₆" hole
⅛" deep

1"

#18 x ⅜" brad

Miter end of Ⓑ at 45°

³⁄₃₂" hole

1¾"

¼" hole

Ⓗ
½" dowel ¼" long

¼"
screw eyes

¾"

¹⁄₁₆" holes
⅛" deep

9"

Ⓒ
¼" dowel
21½" long

EXPLODED VIEW

¾" dia. wooden ball

¼" hole

Bill of Materials			
Part	**Finished Size***		
	T	**W**	**L**
A handle	1"	3¼"	1
B shank	½"	3¼"	1
C rod	¼"	21½"	1
D reel bars	¼"	3"	4
E reel shaft	¼"	3⅝"	1
F crank arm	¼"	2"	1
G* crank hub	½"	¾"	1
H* shaft retainer	½"	¼"	1
I* crank handle	⅜"	1½"	1
J handle pin	⅛"	⅝"	1
K bobber core	⅜"	1½"	1

*Start with longer stock, and cut to finished length in accordance with how-to instructions.

diameter ball in a handscrew clamp, and drill a ¼" hole through it. Glue the ball onto the end of the fishing rod.

Assemble the bobber and hook, referring to the drawings on *page 75*. Epoxy the magnet into the hook. With a marking pen draw a hook on each side of the hook block. Paint the bobber ball red.

Scrollsaw several fish from ¼"-thick stock. Glue a magnet into each where shown. Finish the fish and the rod as desired.

Install four small screw eyes, three spaced evenly along the rod and one on the end where shown. String the rod with nylon or cotton twine about ¹⁄₁₆" in diameter, passing through the screw-eye guides. Tie one end through the hole on the reel shaft.

Just-for-Dowels Drill-Press Jigs

Thread the other end through the hole in the bobber, tie a knot about 1" from the end, then tie the free end to the screw eye in the hook.

Clamped to the drill-press table, this grooved piece of scrapwood holds dowels steady for flat drilling.

Screw the shorter V-groove block to an upright to drill straight down into the end of the dowel.

Drilling holes into the sides and ends of dowels can challenge even the best woodworkers. The problem: trying to hold round stock in position. Here are a couple of simple jigs that will hold your dowels rock-steady.

Both jigs rely on a V-groove that you cut into a piece of ¾×3×20" scrapwood. To cut the V, tilt the blade of your tablesaw to 30° from vertical, and set the cutting depth to ⅜" as shown in the illustration *at right*. Adjust the fence to place the top of the cut on the centerline of the board. Make one pass over the saw blade, turn the board around, and run it through again, creating a 60° V-groove.

Crosscut a 4¼" piece off one end. The longer piece holds your dowels flat on the table, as shown in the photo *above left*. The shorter piece forms the basis for the second jig, which holds dowels vertically for drilling into their ends.

To complete the vertical jig, cut a 1½×3×3" block, clamp it to the back of the 4¼" piece and check to make sure that the V-groove is perpendicular to the base. Then glue and screw the two pieces together as shown in the Vertical Jig drawing. Use the vertical jig as shown in the photo *above right*.

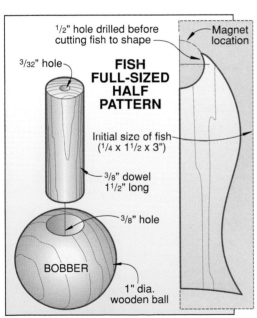

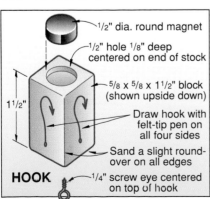

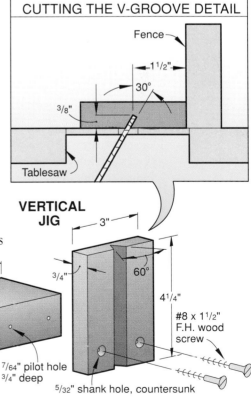

Sunny-Day Sandbox

Hey parents and grandparents, are you looking for the perfect way to keep your kids off the streets and out of trouble? We've got the perfect solution. Our sandbox/toy storage center will keep your kids, and probably many of your neighbors' kids, happily occupied for many fun-filled hours all in the safety of your backyard. And as you can see by looking at the drawings here, this kid-pleasing project is a breeze to build. We completed ours (not including the painting) in just a day. You can, too!

START WITH THE TOYBOX GARAGE

1 From ¾" exterior plywood, cut the front and back (A), ends (B) and bottom (C) to the sizes listed in the Bill of Materials.

2 Cut a ¾" groove ¼" deep ¾" from the bottom edge of the front, back, and end plywood panels (A, B). Cut a ¾" rabbet ½" deep across both ends of the front and back panels.

3 Cut four pieces to ¾×1×21¾" for the corner cleats (D). Drill the countersunk mounting holes through the cleats. (We set up a fence and drilled the holes with a combination countersink bit on our

drill press.) It's easier to drill the holes now than when the cleats are glued in place.

4 Using a caulking gun, apply multi-purpose adhesive to the mating surfaces. Screw the toybox garage (A, B, C, D) together, checking for square.

5 Cut the lid (E) to size from ¾" exterior plywood. Mark and cut a 1½" radius on each corner of the lid. Now, rout a ³⁄₁₆" round-over along all the top and bottom edges of the lid.

6 So they don't protrude, grind or file two corners of a 1½×47" section of the brass continuous (piano) hinge. See the Toybox Garage drawing for reference.

continued

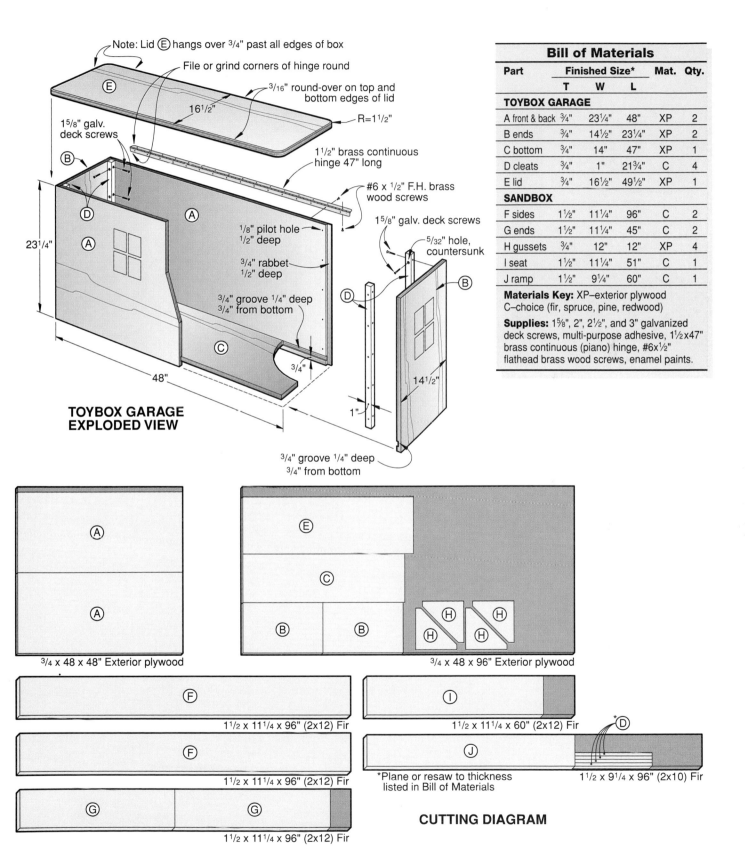

Bill of Materials

Part	Finished Size*			Mat.	Qty.
	T	**W**	**L**		
TOYBOX GARAGE					
A front & back	¾"	23¼"	48"	XP	2
B ends	¾"	14½"	23¼"	XP	2
C bottom	¾"	14"	47"	XP	1
D cleats	¾"	1"	21¾"	C	4
E lid	¾"	16½"	49½"	XP	1
SANDBOX					
F sides	1½"	11¼"	96"	C	2
G ends	1½"	11¼"	45"	C	2
H gussets	¾"	12"	12"	XP	4
I seat	1½"	11¼"	51"	C	1
J ramp	1½"	9¼"	60"	C	1

Materials Key: XP–exterior plywood
C–choice (fir, spruce, pine, redwood)

Supplies: 1⅝", 2", 2½", and 3" galvanized deck screws, multi-purpose adhesive, 1½x47" brass continuous (piano) hinge, #6x½" flathead brass wood screws, enamel paints.

Note: Lid Ⓔ hangs over ¾" past all edges of box

File or grind corners of hinge round

³⁄₁₆" round-over on top and bottom edges of lid

16½"

R=1½"

1⅝" galv. deck screws

1½" brass continuous hinge 47" long

#6 x ½" F.H. brass wood screws

1⅛" pilot hole ½" deep

1⅝" galv. deck screws

23¼"

¾" rabbet ½" deep

⁵⁄₃₂" hole, countersunk

¾" groove ¼" deep ¾" from bottom

¾"

48"

14½"

1"

TOYBOX GARAGE EXPLODED VIEW

¾" groove ¼" deep ¾" from bottom

¾ x 48 x 48" Exterior plywood

¾ x 48 x 96" Exterior plywood

1½ x 11¼ x 96" (2x12) Fir

1½ x 11¼ x 96" (2x12) Fir

1½ x 11¼ x 96" (2x12) Fir

1½ x 11¼ x 60" (2x12) Fir

*Plane or resaw to thickness listed in Bill of Materials

1½ x 9¼ x 96" (2x10) Fir

CUTTING DIAGRAM

Sunny-Day Sandbox

continued

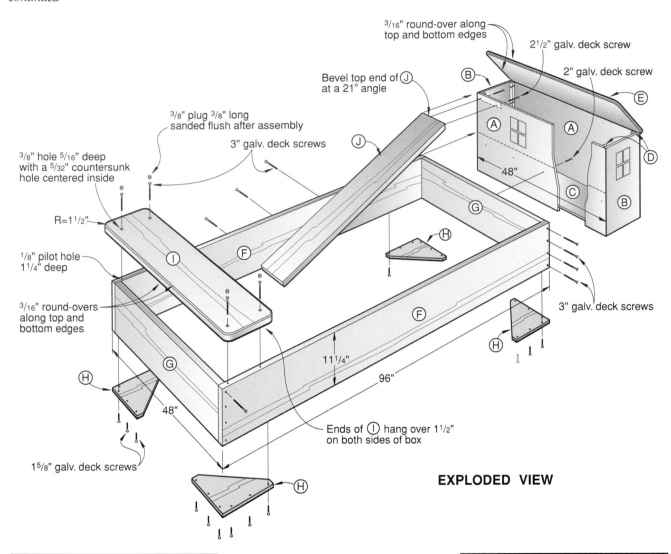

3/16" round-over along top and bottom edges

2½" galv. deck screw

2" galv. deck screw

Bevel top end of Ⓙ at a 21° angle

3/8" plug 3/8" long sanded flush after assembly

3" galv. deck screws

3/8" hole 5/16" deep with a 5/32" countersunk hole centered inside

R=1½"

1/8" pilot hole 1¼" deep

3/16" round-overs along top and bottom edges

48"

11¼"

96"

48"

3" galv. deck screws

Ends of Ⓘ hang over 1½" on both sides of box

1⅝" galv. deck screws

EXPLODED VIEW

A BASIC BOX HOLDS THE SAND

1 From 2×12 stock (we used douglas fir) crosscut the sandbox sides (F) and ends (G) to length.

2 Drill pilot holes, and glue and screw the sandbox pieces (F, G) together, checking for square.

3 To form the gussets (H) to reinforce the corners of the 2×12 box, cut four pieces of ¾"-thick exterior plywood to the shape shown on the Part View drawing.

4 Glue and screw the gussets to the bottom side of the box at the four corners.

5 Cut the seat (I) to length from 2×12" stock. Use a compass to mark a 1½" radius on each corner of the seat. Cut and sand the seat corners to shape. Rout a 3/16" round-over along all the top and bottom edge of the seat.

6 Drill counterbored mounting holes to the sizes listed on the Exploded View drawing. Position, then glue and screw the seat to the box. Cut plugs, plug the seat holes, and sand the plugs flush.

7 Cut the ramp (J) to length from a 2×10, cutting one end at 21° where shown on the Exploded View drawing.

FINAL TOUCHES BEFORE ORDERING THE SAND

1 Sand the sandbox and toybox. Fill any imperfections (we used Minwax High Performance Wood Filler). Sand the filled areas.

2 Apply a coat of primer to both assemblies (don't forget to paint the inside of the toybox garage). (See *opposite page* for painting particulars.) Paint the sandbox and toybox. (We used ACE Porch and Deck enamel).

3 For an added effect, detail-paint the windows on the toybox garage and the logo on the sandbox seat. For the seat logo, we used 5" stencil lettering.

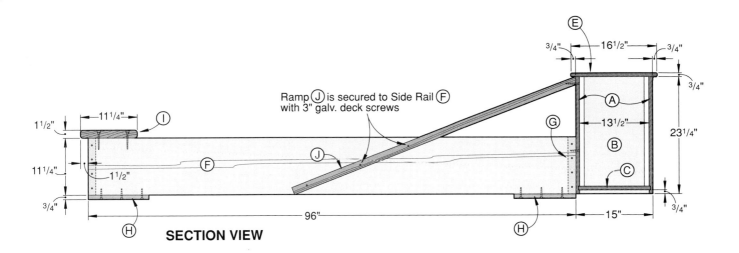

Ramp ⓙ is secured to Side Rail ⓕ
with 3" galv. deck screws

SECTION VIEW

3/4"
16 1/2"
3/4"
3/4"
Ⓔ
3/4"
11 1/4"
Ⓘ
1 1/2"
Ⓐ
13 1/2"
23 1/4"
Ⓖ
Ⓑ
11 1/4"
Ⓕ
ⓙ
Ⓒ
1 1/2"
3/4"
Ⓗ
96"
Ⓗ
15"
3/4"

4 Drill the screw holes and screw
the toybox garage to the sandbox.
5 Position the lid on the toybox,
and then clamp the continuous
hinge in place. Screw the hinge to
the bottom side of the lid and to the
back side of the toybox.
6 Position and then screw the
painted ramp in place. (We used
galvanized screws to fasten the ramp
to the garage front and to the side of
the sandbox.)

HOW WE PAINTED OUR SANDBOX
We first primed the toybox garage
and sandbox pieces with one coat of
an oil-based primer. We then applied
two coats of an industrial oil-based
enamel, letting each coat dry at least
24 hours. For the lettering and
windows, we used an enamel paint.
To paint the windows, we masked
the outlines with masking tape, and
removed the tape within an hour
after painting. Left on too long, the
tape will leave a sticky residue that's
hard to remove.

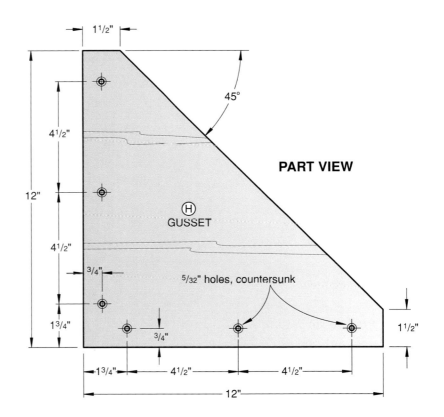

1 1/2"
45°
4 1/2"
PART VIEW
12"
4 1/2"
Ⓗ
GUSSET
4 1/2"
3/4"
5/32" holes, countersunk
1 3/4"
1 1/2"
3/4"
1 3/4"
4 1/2"
4 1/2"
12"

Victory-Lap Clothes Rack

This project is a real winner

LET'S BEGIN WITH THE BACKBOARD AND PEGS

1 From ¾" cherry, cut the backboard to 4×16". Drill a pair of ¼" holes 1" deep into the top edge of the backboard where shown on the Exploded View drawing.

2 To form the checkerboard front for the backboard, cut one piece of walnut and one piece of maple to ¼×2×18". Crosscut eight 2"-long squares from each strip.

3 Draw centerlines on the front of the backboard, and then position, glue, and clamp the 2" maple and walnut squares onto the cherry. After the glues dries, sand the edges of the squares and backboard flush.

4 Angle-cut one end of a piece of scrap at 15°. Fit your portable drill with a ½" brad-point bit, rest the bit against the angled end of the scrap-stock guide, and drill the three ¾"-deep peg holes in the front of the backboard where shown on the Exploded View drawing.

5 From ½" walnut dowel stock, crosscut three pieces to 3⅜" long for the pegs. Sand a ¹⁄₁₆" chamfer on one end of each. Glue the opposite end into the ½" holes in the backboard assembly.

NOW, BUILD THE FORMULA RACE CAR

1 Make two photocopies of the full-sized pattern. Transfer one pattern to ¾" cherry, and cut the car body to shape. Then, transfer the patterns for the front and rear spoilers, fuel tank, helmet/visor, and windshield to the species and thickness specified for each on the full-sized pattern. Next, scrollsaw the pieces to shape. Note that the fuel tank is bevel-ripped at 18° along the top edge, and bevel-cut 63° across the front end.

2 For mounting the coat rack to the wall later, drill a pair of ⁵⁄₃₂" holes through the car body where shown.

3 Sand the pieces smooth, and glue and clamp the pieces you just cut to the car body.

4 Use dowel centers to transfer the centerpoints from the two ¼" holes in the top of the backboard to the bottom of the car body. Drill the 1"-deep mounting holes in the body. Later, sand the edges of the rear

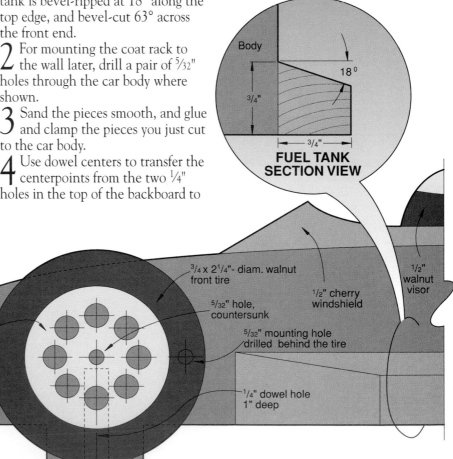

Body

18°

¾"

¾"

FUEL TANK SECTION VIEW

¾ x 2¼"- diam. walnut front tire

⁵⁄₃₂" hole, countersunk

½" cherry windshield

½" walnut visor

⁵⁄₃₂" mounting hole drilled behind the tire

¼" dowel hole 1" deep

½ x 1½"- diam. maple rim

¾" cherry front spoiler

spoiler flush with those of the car body.

5 Cut a pair of 1⅞"-long dowels from ¼" dowel stock. Glue and dowel the car body assembly to the checkerboard backboard.

THE RIMS AND TIRES COME NEXT

1 To form the rims, transfer the rim pattern to ½" maple, drill the ¼" holes, switch bits, and drill and countersink a 5/32" shank hole through the center of each marked rim. Scrollsaw the rims to shape. (We found it safer to drill and then cut the rims to shape).

2 Next, adhere the full-sized tire patterns to ¾" walnut stock. Use a Forstner bit to bore a 1½"-diameter hole centered in each tire. Then, cut the tires to shape.

3 Sand or rout a ⅛" round-over along the front inside and outside edges of each tire. (We used double-faced tape to adhere the tires to the top of our workbench, and then routed the edges.)

4 Check the fit of the maple rims inside the walnut tires. Sand if necessary for a snug fit. Glue the

rims inside the tires, keeping the back surfaces flush. Drill 7/64" pilot holes in the body, and use a #8×1" brass wood screw to fasten each tire/rim to the body.

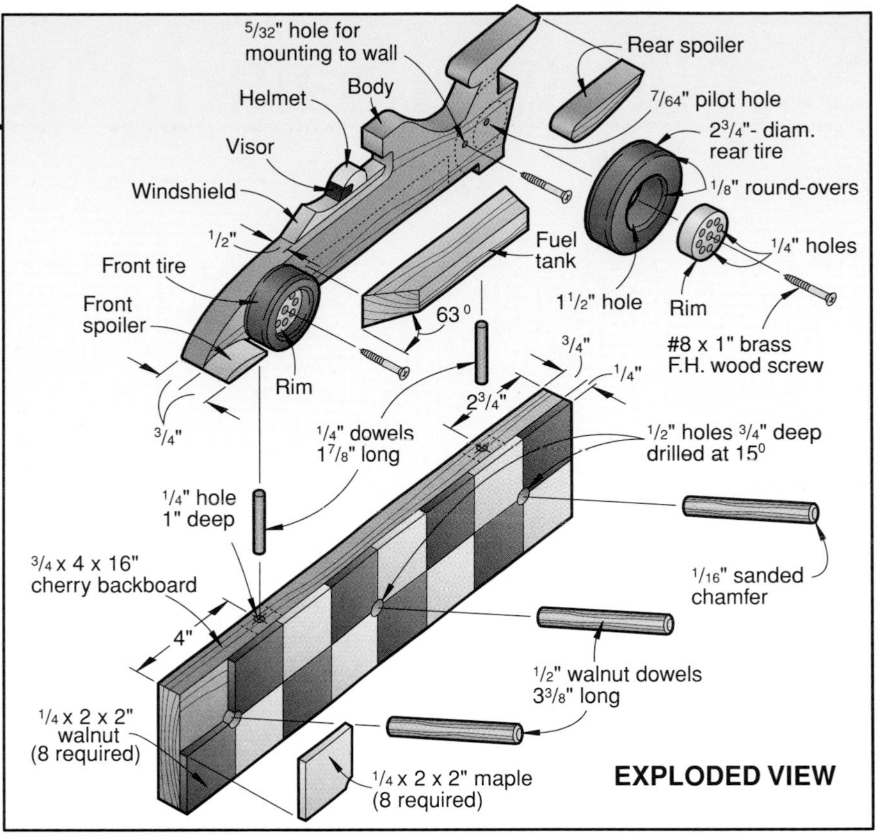

FINISH AND HANG YOUR RACK

1 Finish-sand the car body and backboard. Apply a clear finish (we used an aerosol lacquer).

2 Remove the tire/rim assemblies, and use the

previously drilled holes through the car body to secure the clothes rack to the wall (we used toggle bolts for hanging on drywall). Screw the tire/rims back in place to hide the mounting screws or toggle bolts.

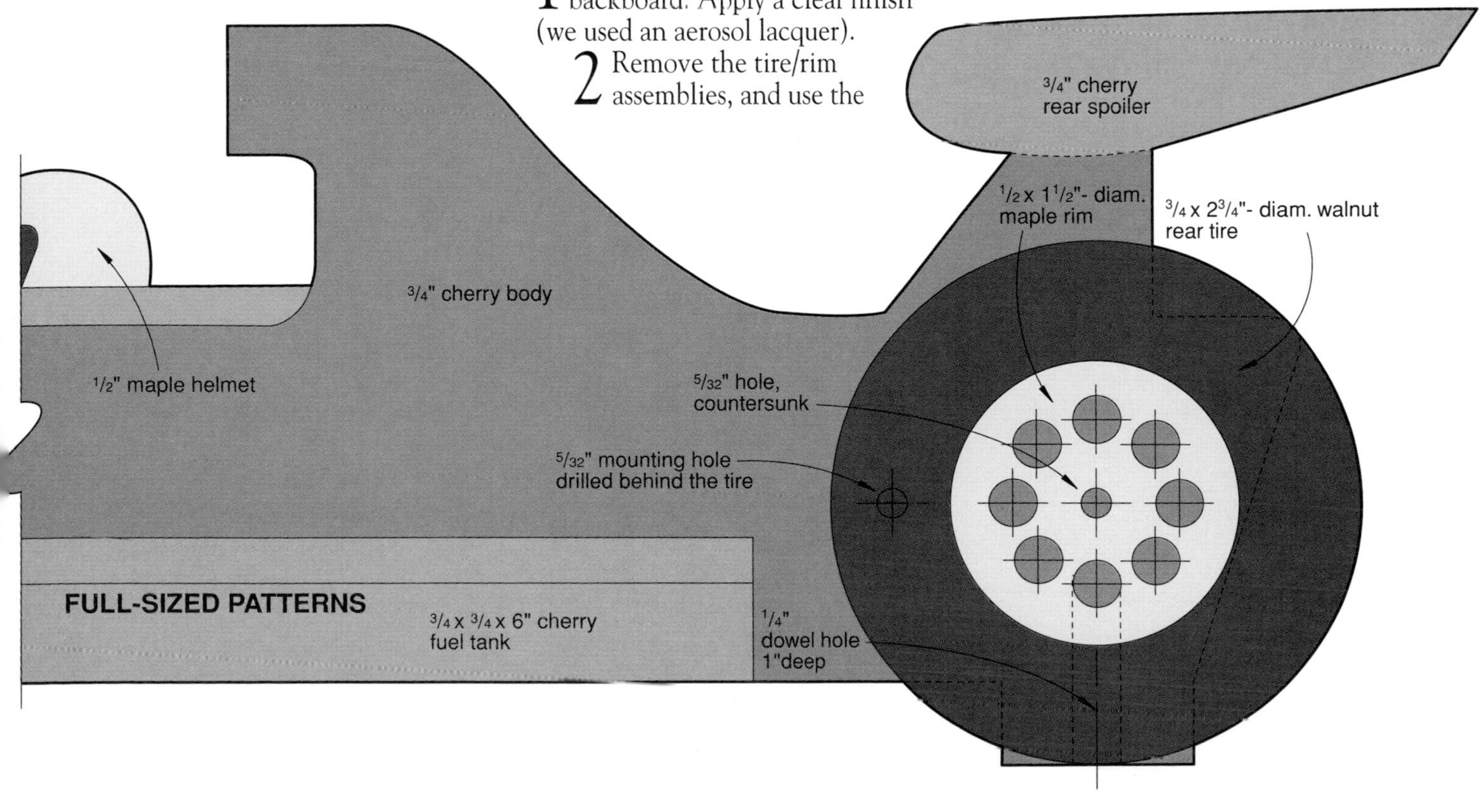

EXPLODED VIEW

FULL-SIZED PATTERNS

Scrollsawed Safari Puzzle

Join us for this exciting scrollsaw tour through the animal kingdom. Wherever you look on this wild pattern, you'll spot a different animal nestled among friends. After cutting out this challenging design, you'll feel like the King of the Scrollsaw Jungle.

Note: *You'll need two 8½×10" pieces of ¼" birch plywood.*

Photocopy the full-sized pattern on the *opposite page*. Spray a light coating of mounting adhesive (we used 3M 77 Spray Adhesive) on the back of the pattern, and allow it to dry for a minute or so. When the glue becomes tacky, center the pattern on the good side of one piece of birch plywood.

Drill ¹⁄₁₆" holes where indicated by the *red* dots on the pattern. Thread the blade through the starting hole on the elephant. (We used a #5 blade, .037×.015", with 14 teeth per inch.)

Scrollsaw the interior cuts designated by red pattern lines first. Making the sharp inside cuts calls for spinning the workpiece around the blade as you cut.

To do this, stand directly in front of the saw, and feed the workpiece steadily into the blade. When you reach a sharp turn, reduce the feeding pressure and simultaneously spin the workpiece sharply, pivoting on the blade. Feed the work into the blade again when the pattern line aligns with the front of the blade. Avoid putting sideways pressure on the blade.

The hole marked "Last" acts as the blade-start hole for sawing around the perimeter of the puzzle. After threading the blade through the

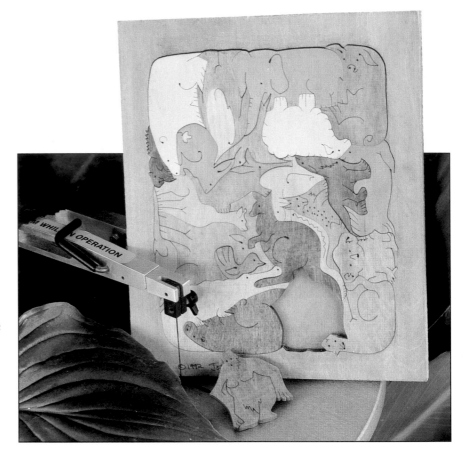

hole, saw along the red line in the direction indicated by the arrow.

When you reach the *blue* perimeter line, saw along it to free the center of the puzzle from the outer frame. Lift the center out, then glue the outer frame to the other piece of plywood.

Now, cut out the animals. From the point marked "Begin animals," cut around the polar-bear, sawing the short cuts for the mouth and toes as you come to them. Cut out the rest of the animals the same way, placing each completed one inside the frame.

Sand as necessary, and finish the puzzle with transparent colors. (We used Delta Home Decor pickling gels

and gel stains and Delta Ceramcoat acrylic colors, all available from craft-supply dealers.) Colors shown (pickling gels and gel stains unless specified) are: BL–Navy blue; BU–Burgundy; CG–Cactus green; CH–Cherry; DW–Driftwood; GO–Goldenrod; MG–Mahogany; ML–Maple; MW-Modern walnut; OR-Ceram-coat orange mixed 50-50 with Home Decor neutral gel; RP–Rose pink; SO–Sunset orange;WH–Desert white. Stain the back and frame with a 50-50 mix of Delta Ceramcoat Black green and Home Decor neutral gel. Wipe on Desert white and wipe it off again to glaze the frame and back.

FULL-SIZED PATTERN

Pickup from the Past

It's an oldie but a goodie

Bill Trumble of Grants Pass, Oregon, designed and built this flashy little truck specifically for our Build-A-Toy™ contest. He did such an outstanding job he won $300 worth of Behlen finishes for his unselfish efforts and originality. A retired teacher, Bill applied paste food colors (the same as his wife uses in her baking) to his truck. We found acrylic paints easier to apply and less likely to run. See the Buying Guide at the end of the article for our source of paints.

Note: *We used thin stock for this project. You can resaw or plane thicker stock to the thicknesses listed in the Bill of Materials.*

BUILD UP FROM THE CHASSIS

1 Cut the chassis (A) to the size listed in the Bill of Materials.

2 Mark and drill the four ⅛" axle holes ⅝" deep in the chassis where dimensioned on the Side View drawing below.

3 Plane or resaw thicker stock to obtain ¼"-thick pine for the cab sides (B) and cab back (C). Using a photocopy and spray adhesive, transfer the full-sized patterns for the parts from *page 87* insert in the center of the magazine to the ¼" stock. Scrollsaw or bandsaw the pieces to shape. Then, drill blade start holes and cut the window openings to shape.

4 Cut the roof blank (D) to the size listed in the Bill of Materials. Transfer the full-sized roof side-view pattern to the blank and cut it to shape. (As shown in the photo *below right*, we used double-faced tape to adhere a piece of scrap 2×4 stock to the roof blank to act as a handle when cutting the roof to shape. We found this much safer than trying to cut the piece without support.) Check that the top of the cab sides (B) mate flush with the curved bottom surface of the roof (D).

Note: *To finish the truck as seen in the photo above left, paint the pieces (except on the surfaces where they'll be glued to another piece) with acrylic paints before assembling them. When we painted the pieces with acrylic paints, we masked the mating areas with masking tape before painting. Then, we glued the pieces together with woodworker's glue. You also can paint all the surfaces of the pieces and glue the parts with instant glue (cyanoacrylate).*

continued

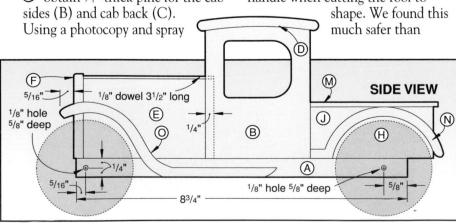

SIDE VIEW

⁵/₁₆"
⅛" dowel 3½" long
⅛" hole ⅝" deep
¼"
¼"
⁵/₁₆"
8¾"
⅛" hole ⅝" deep
⅝"

F M N J H E O B A D

Use a scrap block to support the roof blank when cutting the roof to shape.

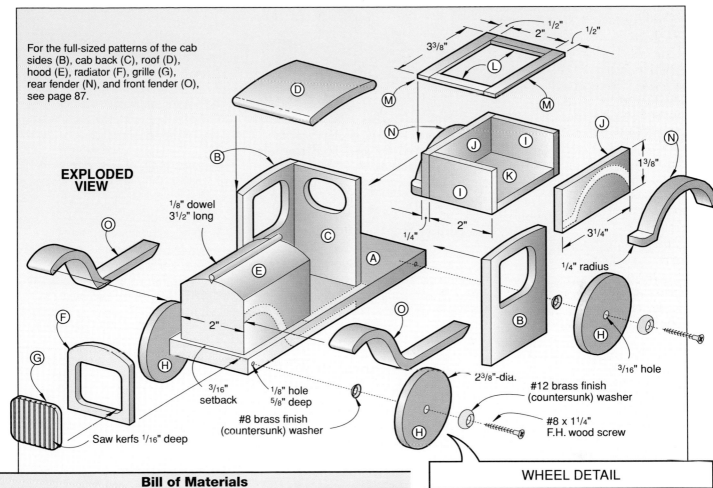

For the full-sized patterns of the cab sides (B), cab back (C), roof (D), hood (E), radiator (F), grille (G), rear fender (N), and front fender (O), see page 87.

EXPLODED VIEW

1/8" dowel 3 1/2" long

3/16" setback

1/8" hole 5/8" deep

Saw kerfs 1/16" deep

#8 brass finish (countersunk) washer

2" 1/2" 1/2"
3 3/8"

1/4"

2"

1 3/8"

3 1/4"

1/4" radius

3/16" hole

2 3/8"-dia.

#12 brass finish (countersunk) washer

#8 x 1 1/4" F.H. wood screw

Bill of Materials

Part	Finished Size*			Mat.	Qty.
	T	W	L		
A chassis	1/2"	2 1/2"	8 3/4"	P	1
B cab sides	1/4"	2 3/4"	3 1/2"	P	2
C cab back	1/4"	2"	3 3/8"	P	1
D* roof blank	3/4"	2 1/2"	3 1/4"	P	1
E hood	2"	2 1/8"	3 1/2"	LP	1
F radiator	1/4"	2"	2 1/4"	P	1
G grille	3/16"	1 1/2"	1 3/8"	P	1
H wheels	3/8"	2 3/8"-diam.		P	4
I box front & back	1/4"	1 3/8"	2"	P	2
J box sides	1/4"	1 3/8"	3 1/4"	P	2

Part	Finished Size*			Mat.	Qty.
	T	W	L		
K box bottom	1/4"	2"	2 3/4"	P	1
L trim	1/8"	1/2"	2"	P	2
M trim	1/8"	1/2"	3 3/8"	P	2
N* rear fender blank	1/2"	1 1/2"	8"	P	1
O* front fender blank	3/4"	2 1/4"	12"	P	1

* Initial size

Materials Key: P–pine, LP–laminated pine

Supplies: 4–#8 brass finish (countersunk) washers, 4–#12 brass finish (countersunk) washers, 4–#8x1 1/4" flathead brass wood screws, 1/8" dowel stock, acrylic paints (listed in Buying Guide at the end of the article).

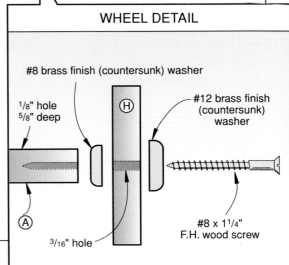

WHEEL DETAIL

#8 brass finish (countersunk) washer

1/8" hole 5/8" deep

#12 brass finish (countersunk) washer

3/16" hole

#8 x 1 1/4" F.H. wood screw

Pickup from the Past

continued

5 Laminate thinner stock to form the 2"-thick hood (E). Trim to length, transfer the pattern to the front end, and bandsaw or sand the top of the hood to shape. Cut a slight groove down the top center of the hood. Bandsaw a piece of ⅛" dowel to the same length as the hood. Paint the dowel and glue it in place in the groove.

6 Center the hood (E) side-to-side on the chassis and ³⁄₁₆" back from the front edge. Glue and clamp the hood in place. Later, glue and clamp the cab pieces (B, C) and roof (D) in place.

7 Cut the radiator (F) and grille (G) to shape. Use a scrollsaw or coping saw to cut the grille opening in the radiator.

8 To highlight the grille use a dovetail saw to cut kerfs ¹⁄₁₆"-deep in it. Then, center and glue the radiator to the front end of the hood. Glue the grille in the opening in the radiator.

THE WHEELS LET 'ER ROLL

1 Using a bandsaw or scrollsaw, cut the four wheels (H) to shape from ⅜"-thick stock. Drill a ³⁄₁₆" hole centered in each wheel.

2 Paint the wheels red. Later, chuck a ³⁄₁₆×2" bolt with a wheel attached to it into your drill press.

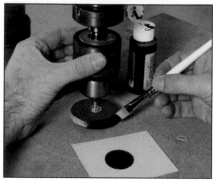

Secure a wheel on a bolt in your drill press. Turn the chuck by hand to paint the black on the wheels.

Start the drill, and sand the edge if necessary for a perfectly round wheel. Stop the drill and rotate the arbor by hand to paint the black rims as shown in the photo above. Repeat for each wheel.

3 Referring to the Wheel detail, use #8×1¼" flathead brass wood screws and finish (countersunk) brass washers to secure the wheels to the chassis. (To keep the screws from working loose over time, we added a drop of instant glue to each screw pilot hole, and immediately drove the screw until the wheel turned freely without wobbling.)

ADD A BOX FOR PLENTY OF PAYLOAD

1 Cut the truck box front and back (I), sides (J), bottom (K), and trim pieces (L, M) to size.

2 Glue and clamp the box pieces (I, J, K), being careful not to scratch the soft pine, and scrape off the excess glue. Then, glue the trim pieces (L, M) in place.

3 Center and glue the box to the chassis directly behind the cab. The outside surfaces of the box should be flush with the outside edges of the chassis.

FENDERS KEEP DOWN THE MUD

1 Cut the rear fender blank (N) and front fender blank (O) to the sizes listed in the Bill of Materials.

2 Transfer the full-sized side view rear fender pattern to the outside face of the rear fender blank (we used two photocopies). Bandsaw the rear fenders to shape (we used a ⅛" blade).

3 To form the front fenders, cut a ¼" rabbet ½" deep along the bottom inside edge of the fender blank where shown on the Parts View drawing. Next, transfer the side view full-sized patterns to the front fender blank. Note that you'll have to make a pair of patterns; one for the right side and one for the left side. Cut along the side view pattern to cut the fenders to shape.

4 Finish-sand the front and rear fenders. Paint the fenders, then glue and clamp them to the chassis, hood, and truck box where shown on the Side View and Exploded View drawings.

BUYING GUIDE

• **Paint.** Two-ounce bottles of Delta Ceramcoat crimson 02076 (red), yellow 02504, copen blue 02051, and black 02506. $1.99/bottle, plus $3.50 postage and handling per order. Meisel Hardware Specialties, P.O. Box 70W, Mound, MN 55364. Credit card orders, call 800/441-9870.

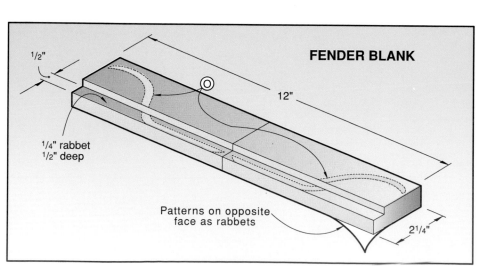

1/2"

FENDER BLANK

12"

1/4" rabbet
1/2" deep

Patterns on opposite
face as rabbets

2¹⁄₄"

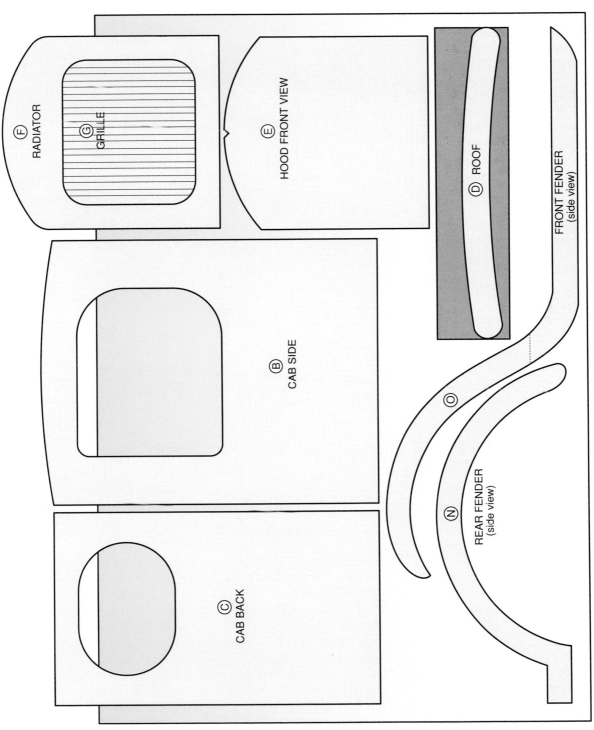

FULL-SIZED PATTERNS

Turn-of-the-Century Snow Stroller

Your little one(s) will love it!

Explore winter's wonderland with your favorite snow angel using this smooth-riding white-oak sled. To push it like a stroller, just flip the handle up and lock it in place. Or, fold the handle down and pull the rope like you do with any other sled, and be on your merry way.

WHY WE USED WHITE OAK FOR THIS PROJECT

Chances are good that this great-looking sled will get lots of use, both now and in the future. To make sure that it will be able to withstand the elements as well as the inevitable wear and tear, we selected white oak. Native to the eastern United States and Canada, white oak rates as one of the strongest, heaviest, and hardest of all the oaks. The pores of this species contain tyloses—bubblelike structures in the vessels of certain hardwoods that help to prevent liquids from penetrating. This attribute makes white oak a perfect material for liquid containers and barrels, and an excellent candidate for outdoor projects such as our sled.

Although you shouldn't have any trouble locating white oak for the parts, we found white oak dowel stock nearly impossible to find. For this reason, we used round-over bits to machine our own dowel stock as described in the story.

LET'S START WITH THE RUNNERS

1 From ¾"-thick white oak, cut the runners (A) to the size listed in the Bill of Materials.

2 Using double-faced tape, stick the two runner blanks together face-to-face with edges and ends flush. The taped-together surfaces will be the *inside* surfaces of the finished sled. For reference, mark the *outside* surfaces with an X.

3 Lay out the runner pattern on the runner blank using the Parts View and Full-Sized Runner End patterns on *opposite page*. Bandsaw along the marked lines to cut the runners to shape. Sand the cut edges to remove the saw marks.

4 Bore a ½" rope hole through the front end of the taped-together runners where marked.

5 Mark the centerpoints for the five ⁵⁄₃₂" holes on one runner where shown on the Runner portion of the Parts View drawing. Drill ⁵⁄₃₂" holes through both runners at each centerpoint. (Because a twist-drill bit tends to wander with the grain in

white oak, we used brad-point bits to eliminate bit wander.) Countersink the ⁵⁄₃₂" holes on the *outside surfaces*—those marked with an X. Mark the centerpoint for the ³⁄₁₆" hole, and drill it where marked.

6 With the runners still taped together, mark the four centerpoints for the ½" dowel holes centered along the *top edge* of each runner where shown on the Parts View drawing. Drill ½" holes ½" deep where marked.

7 Use a wood wedge to pry the runners apart. Remove the carpet tape and any sticky residue with lacquer thinner. Countersink the

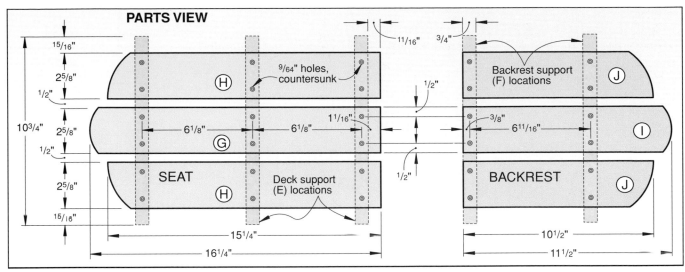

PARTS VIEW

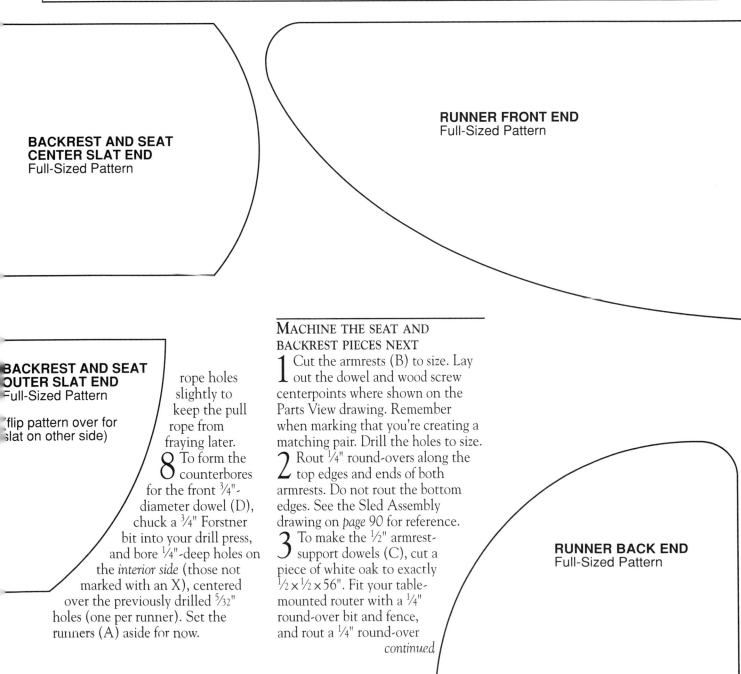

BACKREST AND SEAT CENTER SLAT END
Full-Sized Pattern

RUNNER FRONT END
Full-Sized Pattern

BACKREST AND SEAT OUTER SLAT END
Full-Sized Pattern

(flip pattern over for slat on other side)

RUNNER BACK END
Full-Sized Pattern

MACHINE THE SEAT AND BACKREST PIECES NEXT

1 Cut the armrests (B) to size. Lay out the dowel and wood screw centerpoints where shown on the Parts View drawing. Remember when marking that you're creating a matching pair. Drill the holes to size.

2 Rout ¼" round-overs along the top edges and ends of both armrests. Do not rout the bottom edges. See the Sled Assembly drawing on *page 90* for reference.

3 To make the ½" armrest-support dowels (C), cut a piece of white oak to exactly ½ x ½ x 56". Fit your table-mounted router with a ¼" round-over bit and fence, and rout a ¼" round-over

rope holes slightly to keep the pull rope from fraying later.

8 To form the counterbores for the front ¾"-diameter dowel (D), chuck a ¾" Forstner bit into your drill press, and bore ¼"-deep holes on the *interior side* (those not marked with an X), centered over the previously drilled 5/32" holes (one per runner). Set the runners (A) aside for now.

continued

Turn-of-the-Century Snow Stroller

continued

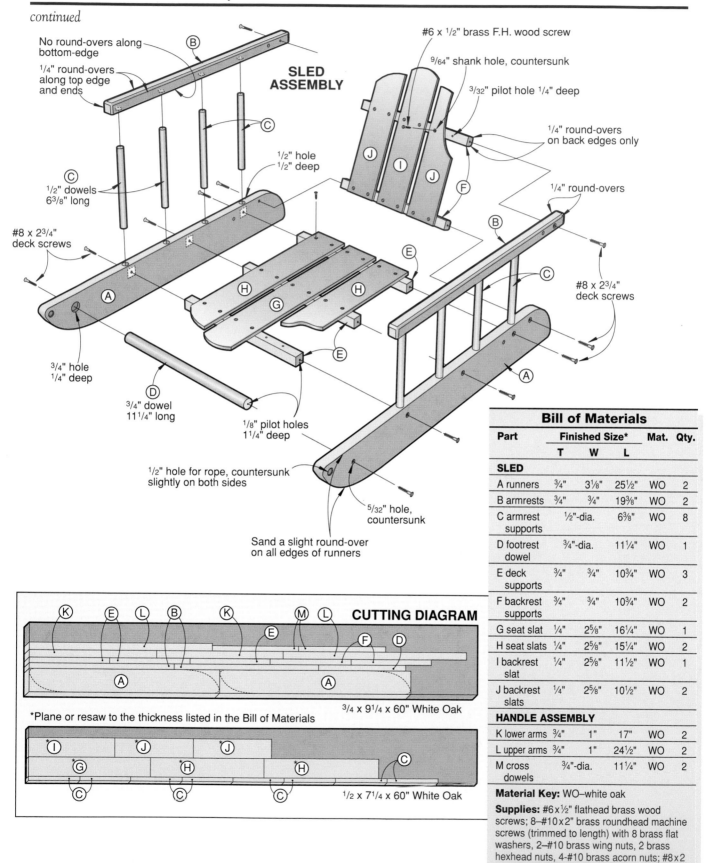

No round-overs along bottom-edge

¹/₄" round-overs along top edge and ends

B

SLED ASSEMBLY

C

C
¹/₂" dowels 6³/₈" long

¹/₂" hole ¹/₂" deep

#8 x 2³/₄" deck screws

A

³/₄" hole ¹/₄" deep

D
³/₄" dowel 11¹/₄" long

¹/₈" pilot holes 11¹/₄" deep

¹/₂" hole for rope, countersunk slightly on both sides

H   G   H

E

E

#6 x ¹/₂" brass F.H. wood screw

⁹/₆₄" shank hole, countersunk

³/₃₂" pilot hole ¹/₄" deep

¹/₄" round-overs on back edges only

J   I   J

F

B

¹/₄" round-overs

C

#8 x 2³/₄" deck screws

A

⁵/₃₂" hole, countersunk

Sand a slight round-over on all edges of runners

Bill of Materials

Part	Finished Size*			Mat.	Qty.
	T	W	L		
SLED					
A runners	³/₄"	3¹/₈"	25¹/₂"	WO	2
B armrests	³/₄"	³/₄"	19³/₈"	WO	2
C armrest supports	¹/₂"-dia.		6³/₈"	WO	8
D footrest dowel	³/₄"-dia.		11¹/₄"	WO	1
E deck supports	³/₄"	³/₄"	10³/₄"	WO	3
F backrest supports	³/₄"	³/₄"	10³/₄"	WO	2
G seat slat	¹/₄"	2⁵/₈"	16¹/₄"	WO	1
H seat slats	¹/₄"	2⁵/₈"	15¹/₄"	WO	2
I backrest slat	¹/₄"	2⁵/₈"	11¹/₂"	WO	1
J backrest slats	¹/₄"	2⁵/₈"	10¹/₂"	WO	2
HANDLE ASSEMBLY					
K lower arms	³/₄"	1"	17"	WO	2
L upper arms	³/₄"	1"	24¹/₂"	WO	2
M cross dowels	³/₄"-dia.		11¹/₄"	WO	2

Material Key: WO–white oak

Supplies: #6x¹/₂" flathead brass wood screws; 8–#10x2" brass roundhead machine screws (trimmed to length) with 8 brass flat washers, 2–#10 brass wing nuts, 2 brass hexhead nuts, 4-#10 brass acorn nuts; #8x2 and #8x2³/₄" deck screws; ³/₈" braided nylon rope 8' long; clear exterior finish.

CUTTING DIAGRAM

K   E   L   B   K   M   L

E

F   D

A   A

³/₄ x 9¹/₄ x 60" White Oak

*Plane or resaw to the thickness listed in the Bill of Materials

I   J   J

G   H   H   C

C   C   C

¹/₂ x 7¹/₄ x 60" White Oak

along all four edges of the 56"-long piece. Sand the dowel smooth.

4 Crosscut eight pieces of 6⅜"-long dowel stock from the 56"-long piece.

5 To create the ¾"-diameter footrest dowel (D), repeat the process in Step 3 using ¾"-square white oak 40" long and a ⅜" round-over bit. Crosscut a 11¼"-long piece from the stock for the footrest dowel. Set the other piece of dowel aside; we'll use it later for the handle dowels (M).

6 Cut the deck and backrest supports (E, F) to size. Mark diagonals on both ends of each support to find center, and drill a ⅛" pilot hole at each marked centerpoint.

7 Rout ¼" round-overs along the back edge of each rear support (F) where shown on the Sled Assembly drawing.

NOW, ADD THE SLATS AND ARMRESTS

1 From ¼"-thick stock, cut the seat and backrest slats (G, H, I, J) to size. (We planed thicker stock to the required ¼" thickness.) Using the Full-Sized End patterns and the Parts View on *page 89*, transfer the patterns to the slats. (We transferred the shapes to hardboard, cut these to shape, and used them as templates to mark all the slats.) Cut and sand the slat ends to shape.

2 Mark the hole centerpoints shown on the Seat and Backrest Parts View drawing onto the slats. Drill the countersunk screw holes where marked.

3 Glue and clamp the ¾" oak footrest dowel (D), and deck supports (E) between the runners (A). Drive the screws through the previously drilled shank and pilot holes. (To avoid twisting off the screw, we rubbed the threaded end against a bar of paraffin before driving the screws.)

4 As shown in the photo *above right*, clamp and screw the seat

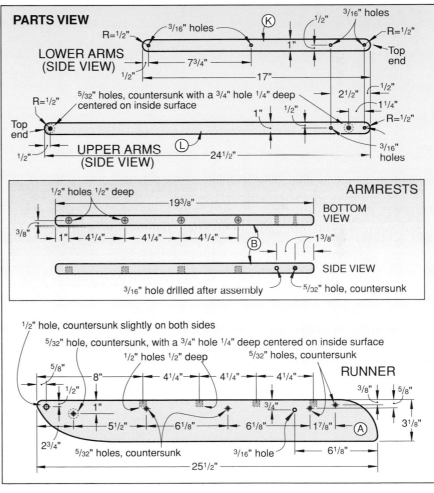

center slat (G) in place. Then, add the side slats (H).

5 Using Titebond II or Weather-Tite water-resistant glue or slow-set epoxy, glue the ½" white oak

dowels (C) between the armrests (B) and runners (A).

6 Screw (don't glue) the backrest supports (F) between the

continued

Turn-of-the-Century Snow Stroller

continued

armrests and runners. Angle the backrest supports so the front edges are parallel. Then, using the same procedure as used in Step 4 on the *previous page* and the Backrest drawing for dimensions, screw the backrest slats (I, J) to the backrest supports (F).

THE HANDLE ASSEMBLY COMES NEXT

1 From ¾" white oak, cut the lower arm pieces (K) and upper arm pieces (L) to size.

2 Clamp the lower-arm pieces parts edge-to-edge. Do the same for the upper-arm pieces. Next, mark the hole centerpoints on each of the two pairs where shown on the Parts View drawing.

3 Use a compass to mark a ½" radius on each end of each handle arm. Separate the pieces, and cut the arm ends to shape.

4 With a ³⁄₁₆" brad-point bit, drill the machine screw holes through all four handle pieces where marked. Switch to a ⁵⁄₃₂" bit, and drill the remaining two holes in the upper arms where marked.

5 Bore a pair of ¾" holes ¼" deep centered over the ⁵⁄₃₂" holes on the *inside face* of the upper arms (L). Crosscut the upper arm dowels (M) from the ¾"-diameter dowel stock formed earlier.

6 Glue and screw the upper arm assembly (L, M) together, checking for square.

7 Use machine screws to secure the *bottom end* of the lower arms to the runners (parts K to A). Next, use machine screws to connect the upper-arm assembly to the lower arms. (To eliminate protruding screws, we started with 2"-long machine screws and used a hacksaw to trim the ends of the screws flush with the outside surface of the nuts.)

8 Position the handle assembly until the holes in the lower arms (K) align with the *center* of the armrests (B). Now, use the previously drilled ³⁄₁₆" holes in the lower arms as guides to drill mating holes through the armrests. See the Handle detail *below left* for reference.

9 Separate the pieces, and finish-sand the sled, upper-arm assembly, and lower arms. Apply at least two coats of spar varnish. Add a rope. Reassemble the parts, and head for the snow.

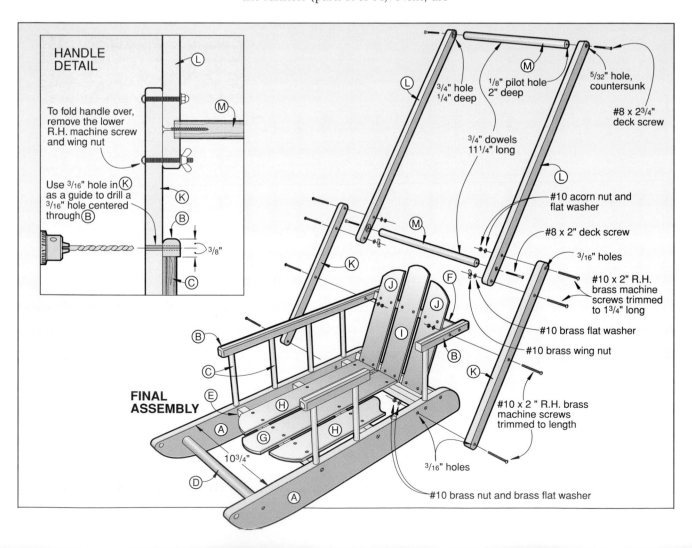

HANDLE DETAIL

To fold handle over, remove the lower R.H. machine screw and wing nut

Use ³⁄₁₆" hole in (K) as a guide to drill a ³⁄₁₆" hole centered through (B)

3/8"

¾" hole ¼" deep

1/8" pilot hole 2" deep

⁵⁄₃₂" hole, countersunk

#8 x 2¾" deck screw

¾" dowels 11¼" long

#10 acorn nut and flat washer

#8 x 2" deck screw

³⁄₁₆" holes

#10 x 2" R.H. brass machine screws trimmed to 1¾" long

#10 brass flat washer

#10 brass wing nut

#10 x 2 " R.H. brass machine screws trimmed to length

FINAL ASSEMBLY

10¾"

³⁄₁₆" holes

#10 brass nut and brass flat washer

The Fun-Time Racer

An adventure just waiting to happen

Have you ever stumbled onto a woodworking project whose looks and styling stopped you cold? Well, that's what happened to WOOD® magazine design editor Jim

Downing and me not long ago. We were passing by a local antique store, and there it was—the neatest old kid's wagon in creation. We looked at each other and decided

right then that we wanted to design a version of it just for you.

We were careful to retain its classic lines, and to make sure this

continued

The Fun-Time Racer

continued

beauty will stand up to lots of outdoor use. We built the deck and box out of white oak and the wheels from exterior birch plywood. To find the hardware for the wagon, buy the parts at a local hardware store, or use our handy Buying Guide kit. We tell you how to machine these parts to fit.

Marlen Kemmet
How-To Editor

Note: You'll need thin stock for this project. You can resaw or plane thicker stock, or see the Buying Guide with the

Bill of Materials for our source of white oak and exterior birch plywood. Also listed is a supplier for all the hardware and strap iron needed.

LET'S START WITH THE DECK

1 From white oak (one of the heaviest and strongest of all the oaks), cut the deck boards (A) and the cleats (B) to the sizes listed in the Bill of Materials. (We planed ½" stock to ⅜" for the deck boards.) Sand a slight round-over on all edges to avoid splintering.

2 Dry-clamp (no glue) the deck boards edge-to-edge, with the ends flush. Lay out and clamp the cleats to the deck bottom. Turn the

assembly over, and mark the four screw-hole centerlines across the surface of the clamped-together deck boards. See the Exploded View for reference.

3 Position and clamp the cleats (B) on the deck assembly. Drill countersunk screw holes through the deck boards 1" from the edges and ¼" into the cleats. Use #6 × ⅝" screws to secure the cleats to the deck boards. Remove the clamps.

CONSTRUCT THE REAR AXLE ASSEMBLY NEXT

1 From ¾" white oak, cut two pieces to 2½ × 6¼" for the rear

continued

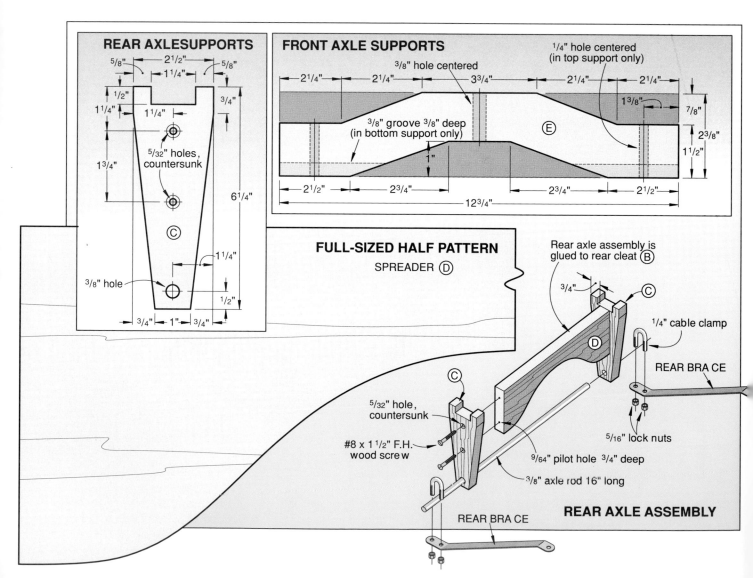

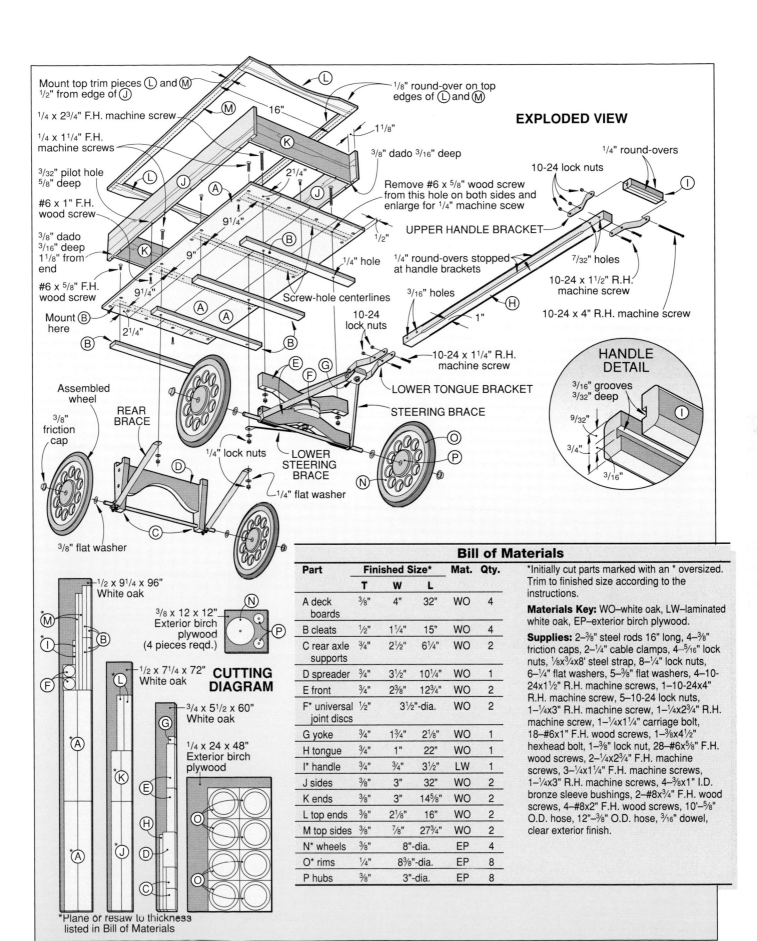

EXPLODED VIEW

Mount top trim pieces Ⓛ and Ⓜ ½" from edge of Ⓙ

¼ x 2¾" F.H. machine screw

¼ x 1¼" F.H. machine screws

³/₃₂" pilot hole ⅝" deep

#6 x 1" F.H. wood screw

³/₈" dado ³/₁₆" deep 1⅛" from end

#6 x ⅝" F.H. wood screw

Mount Ⓑ here

16"

1⅛"

³/₈" dado ³/₁₆" deep

2¼"

9¼"

9"

2¼"

9¼"

¼" hole

Screw-hole centerlines

½"

⅛" round-over on top edges of Ⓛ and Ⓜ

Remove #6 x ⅝" wood screw from this hole on both sides and enlarge for ¼" machine scew

UPPER HANDLE BRACKET

¼" round-overs stopped at handle brackets

³/₁₆" holes

¼" round-overs

10-24 lock nuts

⁷/₃₂" holes

10-24 x 1½" R.H. machine screw

10-24 x 4" R.H. machine screw

10-24 lock nuts

10-24 x 1¼" R.H. machine screw

LOWER TONGUE BRACKET

STEERING BRACE

1"

Assembled wheel

³/₈" friction cap

REAR BRACE

¼" lock nuts

LOWER STEERING BRACE

¼" flat washer

³/₈" flat washer

HANDLE DETAIL

³/₁₆" grooves ³/₃₂" deep

9/32"

¾"

³/₁₆"

CUTTING DIAGRAM

½ x 9¼ x 96" White oak

³/₈ x 12 x 12" Exterior birch plywood (4 pieces reqd.)

½ x 7¼ x 72" White oak

¾ x 5½ x 60" White oak

¼ x 24 x 48" Exterior birch plywood

*Plane or resaw to thickness listed in Bill of Materials

Bill of Materials

Part	Finished Size*			Mat.	Qty.
	T	W	L		
A deck boards	³/₈"	4"	32"	WO	4
B cleats	½"	1¼"	15"	WO	4
C rear axle supports	¾"	2½"	6¼"	WO	2
D spreader	¾"	3½"	10¼"	WO	1
E front	¾"	2⅜"	12¾"	WO	2
F* universal joint discs	½"		3½"-dia.	WO	2
G yoke	¾"	1¾"	2⅛"	WO	1
H tongue	¾"	1"	22"	WO	1
I* handle	¾"	¾"	3½"	LW	1
J sides	³/₈"	3"	32"	WO	2
K ends	³/₈"	3"	14⅝"	WO	2
L top ends	³/₈"	2⅛"	16"	WO	2
M top sides	³/₈"	⅞"	27¾"	WO	2
N* wheels	³/₈"		8"-dia.	EP	4
O* rims	¼"		8⅜"-dia.	EP	8
P hubs	³/₈"		3"-dia.	EP	8

*Initially cut parts marked with an * oversized. Trim to finished size according to the instructions.

Materials Key: WO–white oak, LW–laminated white oak, EP–exterior birch plywood.

Supplies: 2–⅜" steel rods 16" long, 4–⅜" friction caps, 2–¼" cable clamps, 4–⁵/₁₆" lock nuts, ⅛x¾x8' steel strap, 8–¼" lock nuts, 6–¼" flat washers, 5–⅜" flat washers, 4–10-24x1½" R.H. machine screws, 1–10-24x4" R.H. machine screw, 5–10-24 lock nuts, 1–¼x3" R.H. machine screw, 1–¼x2¾" R.H. machine screw, 1–¼x1¼" carriage bolt, 18–#6x1" F.H. wood screws, 1–⅜x4½" hexhead bolt, 1–⅜" lock nut, 28–#6x⅝" F.H. wood screws, 2–¼x2¾" F.H. machine screws, 3–¼x1¼" F.H. machine screws, 1–¼x3" R.H. machine screws, 4–⅜x1" I.D. bronze sleeve bushings, 2–#8x¾" F.H. wood screws, 4–#8x2" F.H. wood screws, 10'–⁵/₈" O.D. hose, 12"–⅜" O.D. hose, ³/₁₆" dowel, clear exterior finish.

The Fun-Time Racer

continued

axle supports (C). Using double-faced tape, stick the two pieces together, with the edges and ends flush. Using the dimensions on the Rear Axle Supports drawing on *page 94*, lay out the shape, notch, and hole centerpoints on the top piece.

2 Drill the countersunk screw and axle holes through the taped-together pieces. Bandsaw the pieces to shape. Check the fit of the notch against the rear cleat on the wagon deck, and trim if necessary. Separate the parts, remove the tape, and sand smooth.

3 Cut a piece of ¾" stock to 3½ × 10¼" for the spreader (D). Transfer the full-sized half pattern twice to the spreader blank, and cut it to shape.

4 Glue and screw the rear axle assembly (C, D) together where shown on the Rear Axle Assembly drawing. (We inserted the axle rod in the holes in C to keep the pieces aligned when assembling.)

ADD THE FRONT STEERING ASSEMBLY

1 Cut a pair of ¾"-thick front axle support blanks (E) to 2⅜ × 12¾". Using the Steering Assembly and Front Axle Supports drawings for reference, drill the ¼" and ⅜" holes in the top support blank. Drill a ⅜" hole centered in the bottom blank. Next, cut a ⅜" groove ⅜" deep along the bottom edge of the bottom blank (E).

2 Transfer the cutlines from the Front Axle Supports drawing to

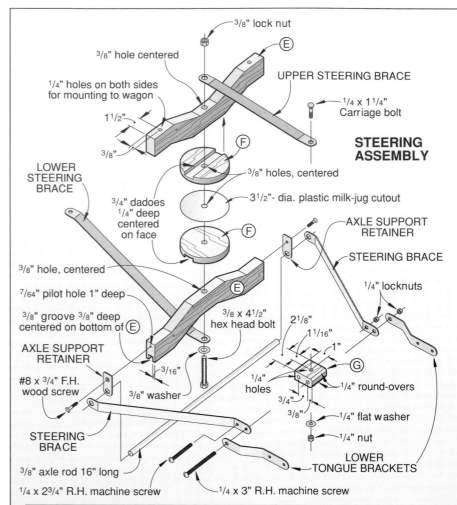

STEERING ASSEMBLY

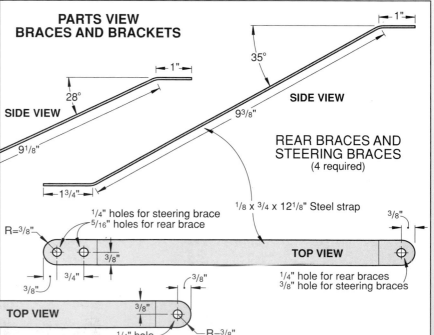

PARTS VIEW
BRACES AND BRACKETS

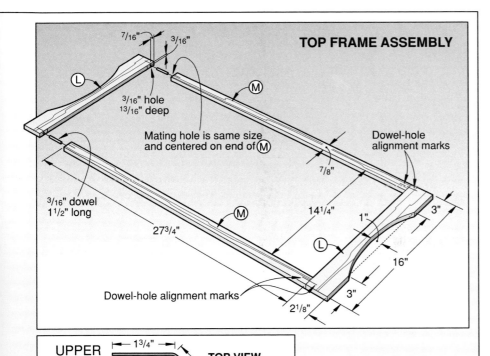

TOP FRAME ASSEMBLY

7/16"

3/16"

Ⓛ Ⓜ

3/16" hole
13/16" deep

Mating hole is same size
and centered on end of Ⓜ

Dowel-hole
alignment marks

7/8"

3/16" dowel
1 1/2" long

Ⓜ

14 1/4"

1"

3"

27 3/4"

Ⓛ

16"

3"

Dowel-hole alignment marks

2 1/8"

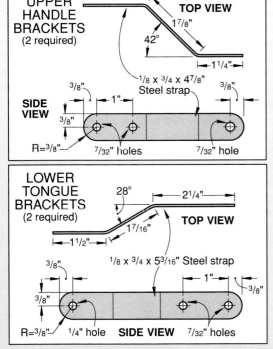

UPPER HANDLE BRACKETS
(2 required)

1 3/4"

TOP VIEW

1 7/8"

42°

1 1/4"

1/8 x 3/4 x 4 7/8"
Steel strap

SIDE VIEW

3/8"   1"   3/8"

3/8"

R=3/8"   7/32" holes   7/32" hole

LOWER TONGUE BRACKETS
(2 required)

28°   2 1/4"

1 7/16"   **TOP VIEW**

1 1/2"

1/8 x 3/4 x 5 3/16" Steel strap

3/8"   1"   3/8"

3/8"

R=3/8"   1/4" hole   **SIDE VIEW**   7/32" holes

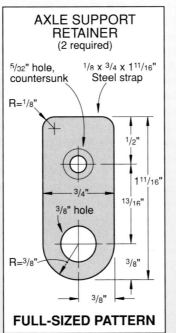

AXLE SUPPORT RETAINER
(2 required)

5/32" hole,
countersunk

1/8 x 3/4 x 1 11/16"
Steel strap

R=1/8"

1/2"

3/4"

1 11/16"

13/16"

3/8" hole

3/8"

R=3/8"

3/8"

FULL-SIZED PATTERN

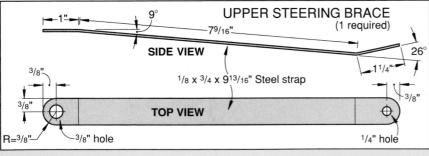

UPPER STEERING BRACE
(1 required)

1"   9°

7 9/16"

SIDE VIEW

26°

1 1/4"

1/8 x 3/4 x 9 13/16" Steel strap

3/8"

3/8"

3/8"   **TOP VIEW**   3/8"

R=3/8"   3/8" hole   1/4" hole

the top E, cut it to shape, and use it as a template to mark the shape on the bottom E. Cut it to shape.

3 To make the universal-joint discs (F), cut a piece of 1/2" stock to 4x12". Mount a dado blade to your tablesaw, and cut a 3/4" dado 1/4" deep centered along one face of the 12"-long piece. Using a compass, mark a pair of 3 1/2"-diameter circles on the surface opposite the dado and centered over the dado. Drill a 3/8" hole at the centerpoint used to mark the circles. Bandsaw and sand the universal-joint discs to shape.

4 Glue and clamp one universal joint disc (F) to each front-axle support (E), using a 3/8 x 4 1/2" hexhead bolt to align the holes.

5 Cut a 3 1/2"-dia. disc from a plastic milk jug for use as a slip joint between the universal-joint discs.

NOW, CONSTRUCT THE YOKE AND HANDLE ASSEMBLY

1 Cut the yoke block (G) to size.

2 Mark the hole centerpoints on the yoke. Clamp the yoke in a handscrew clamp, and drill the holes where marked. (We used a brad-point bit to eliminate bit wander.) Sand 1/4" round-overs along the front edges of the yoke.

3 Cut the tongue (H) to size.

4 To form the handle (I), cut a piece of 3/8" stock to 3/4 x 12". Cut a 3/16" groove 3/32" deep centered along one face of the handle blank. Next, rout 1/4" round-overs along the edges opposite the groove. Trim two pieces to 3 1/2" long from the 12"-long strip. With the edges and grooves aligned, glue and clamp the two pieces together to form the handle.

FORM THE WAGON BOX AND TOP FRAME ASSEMBLIES

1 From 3/8" stock, cut the wagon sides (J) and ends (K) to size.

2 Cut a 3/8" dado 3/16" deep 1 1/8" from the ends in each side piece

continued

The Fun-Time Racer

continued

(J). See the Exploded View drawing for reference.

3 Sand the interior faces of the wagon sides and ends. Glue and clamp the four pieces, checking that the mating edges are flush and that the assembly is square.

4 Cut the top frame members (L, M) to size. Mark a radius on the front and rear pieces (L) where shown on the Top Frame Assembly drawing. Cut the curved sections to shape.

5 Clamp the top frame in the configuration shown on the Top Frame Assembly drawing. Mark the dowel-hole alignment marks across each joint. Remove the clamps, and use a doweling jig to drill $\frac{3}{16}$" holes $\frac{13}{16}$" deep in the mating pieces. Glue, dowel, and clamp the top frame together, checking for square.

6 Center and glue the top frame assembly (L, M) on the wagon box (J, K). (We placed masking tape on the areas next to the glued areas to catch glue squeeze-out.)

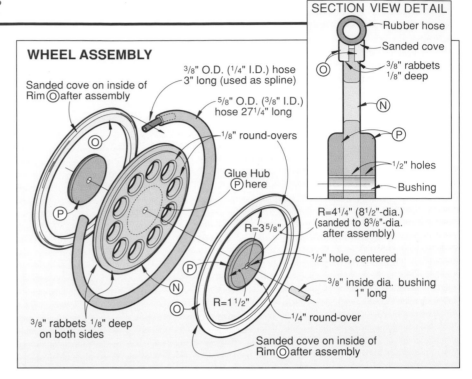

WHEEL ASSEMBLY

Sanded cove on inside of Rim O after assembly

$\frac{3}{8}$" O.D. ($\frac{1}{4}$" I.D.) hose 3" long (used as spline)

$\frac{5}{8}$" O.D. ($\frac{3}{8}$" I.D.) hose $27\frac{1}{4}$" long

$\frac{1}{8}$" round-overs

Glue Hub P here

$\frac{3}{8}$" rabbets $\frac{1}{8}$" deep on both sides

R=$3\frac{5}{8}$"

R=$1\frac{1}{2}$"

R=$4\frac{1}{4}$" ($8\frac{1}{2}$"-dia.) (sanded to $8\frac{3}{8}$"-dia. after assembly)

$\frac{1}{2}$" hole, centered

$\frac{3}{8}$" inside dia. bushing 1" long

$\frac{1}{4}$" round-over

Sanded cove on inside of Rim O after assembly

SECTION VIEW DETAIL

Rubber hose — Sanded cove — $\frac{3}{8}$" rabbets $\frac{1}{8}$" deep — $\frac{1}{2}$" holes — Bushing

7 Sand the inside surfaces where part M meets part J. Rout a $\frac{1}{8}$" round-over along the top edges of the top frame.

8 Carefully position and clamp the wagon box to the deck. Drill countersunk mounting holes, and screw the box to the deck.

FOR A SMOOTH RIDE, ADD THE WHEELS AND TIRES

1 Cut four pieces of $\frac{3}{8}$" exterior plywood to 9" square for the wheel blanks (N). (Because of its smoothness and paintability, we used $\frac{3}{8}$" exterior birch plywood. See the Buying Guide for our source.) Draw diagonal lines to locate the center of each wheel blank. Use a compass to mark an 8"-diameter circle (4" radius) from each marked centerpoint.

2 To lay out ten $1\frac{1}{8}$" hole centerpoints on each

wheel blank (N), reset the compass, and swing a $2\frac{7}{16}$" radius to create a $4\frac{7}{8}$" smaller circle within the 8" circle. See the Wheels drawing for reference. Reset the compass to $1\frac{1}{2}$", start on the $4\frac{7}{8}$" circle, and step off the ten centerpoints. You may have to adjust the compass slightly for exact spacing.

3 Drill a $\frac{1}{8}$" hole at each centerpoint. Switch to a $1\frac{1}{8}$" Forstner bit. One at a time, center the bit over $\frac{1}{8}$" pilot holes, and bore halfway through the wheel. Turn the wheel over, center the bit over the pilot hole, and finish boring the hole. This greatly reduces the chip-out at each hole.

4 Using a $\frac{1}{2}$" brad-point bit (it eliminates bit wander), drill an axle hole centered in each wheel.

5 Cutting just outside the marked line, bandsaw each wheel to shape. Move to the disc sander, and sand the wheels to finished shape. For near perfect-round wheels, see our disc-sanding jig on *page 100*.

6 Fit a $\frac{3}{8}$" rabbeting bit into your table-mounted router, and adjust it to cut $\frac{1}{8}$" deep. Rout a $\frac{3}{8}$" rabbet $\frac{1}{8}$" deep along each edge of the four

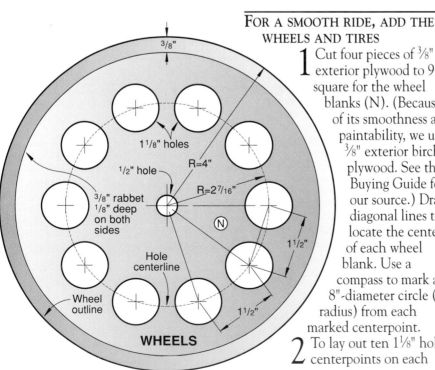

$\frac{3}{8}$"

$1\frac{1}{8}$" holes

R=4"

$\frac{1}{2}$" hole

$\frac{3}{8}$" rabbet $\frac{1}{8}$" deep on both sides

R=$2\frac{7}{16}$"

N

Hole centerline

$1\frac{1}{2}$"

Wheel outline

$1\frac{1}{2}$"

WHEELS

wheels. See the Section View Detail accompanying the Wheel Assembly drawing for reference.

7 Switch to a ⅛" round-over bit, and rout along the edges of the 1⅛" holes in the wheels where shown on the Wheel Assembly drawing *opposite*.

8 To make the rims (O), start by cutting 8 pieces of ¼" exterior birch plywood to 10" square. Using double-faced tape, stick the rim blanks together in pairs. Locate center and use a compass to swing a 3⅝" radius to create a 7¼"-diameter circle on each taped-together pair of blanks. Reset the compass to 4¼", and swing an 8½"-diameter circle on the same four rim blanks.

9 To fit the rims onto the wheels, drill a blade start hole inside each marked inner circle, and use a scrollsaw to cut the rims to shape. Then, use a drum sander to sand the inside edge until the rims fit snugly onto the rabbeted wheels (N). Work slowly to achieve a gap-free fit.

10 Bandsaw and then sand the outside edge of the four sets of rims to shape. With the rims still taped together, rout ⅛" round-overs along the inside edge on one surface of each rim where shown on the Wheel Assembly drawing. Separate the rims and remove the tape.

11 Glue the rims into the rabbets on each wheel. (We wrapped masking tape around the rims and through the 1⅛" holes in the wheels to hold the rims in place until the glue dried.)

12 Using the disc-sanding jig shown on *page 100*, carefully sand the wheel/rim assemblies to 8⅜" diameter. *Verify all are the same diameter when you're done*.

13 Mark the cutlines and then bandsaw and sand the hubs (P) to shape. Drill a ½" hole at the centerpoint of each. Rout a ¼" round-over along the outside edge of each hub. Glue the hubs in place, aligning the holes.

14 Using a hacksaw, cut the head off a ½" bolt 3" long. Chuck the bolt into your drill press, and use two nuts to secure a wheel to it. Using the lowest speed on your drill press, use a rat-tail file to contour the inside edges of the rims to the shape shown on the Section View Detail accompanying the Wheel Assembly drawing and as shown in Photo A. The coves allow the rubber hose (tire) to fit snugly around the outside of the plywood wheel.

A

Use a round file to shape and smooth the inside edge of the wheel rims for a flush fit against the tire.

HOMEMADE TIRES FOR A SOFT, INEXPENSIVE RIDE

1 To make a tire, cut ⅝" O.D. (⅜" I.D.) fuel-vapor hose to 27¼" long. Then, cut four pieces of ⅜" O.D. (¼" I.D.) hose to 3" long.

2 Insert half of the smaller hose into one end of the larger hose where shown on the Wheel Assembly drawing. Slide the opposite end of the larger hose onto the protruding end of the smaller hose. Check for a *tight* fit of the tire onto the wheel. Trim the hose if necessary. Use instant glue (cyanoacrylate) to adhere the mating ends, striving for a tight joint. Repeat Steps 1 and 2 for each tire.

3 Lightly sand the tires to remove any visible lettering or exposed instant glue. Set the tires aside for now; you'll epoxy them to the wheels after finishing.

NEXT, SHAPE AND DRILL THE BRACES AND BRACKETS

1 To form the rear braces and steering braces, start by cutting four pieces of ⅛ × ¾" strap iron to 12⅛" long. Place two pieces of the strapping edge-to-edge (not face-to-face), and adhere duct tape to the back surface to hold them together. Apply masking tape to the opposite surface on the ends. The masking tape makes it easy to mark visible bend lines and radii.

2 Lay out the hole centerpoints and bend lines for the braces on the masking tape where shown on the Braces and Brackets Parts View drawing. Dimple each centerpoint with a center punch. Drill ¼" holes at each dimpled centerpoint, except for the steering braces. They have a ⅜" hole.

3 To bend the ends to the angles shown, clamp the taped-together parts in a machinist's or woodworker's vise so that the marked bend line is flush with the top surface of the vise jaws. As shown in Photo B, use a hammer to strike the steel straps until they conform to the angle shown on the Parts View drawing. Check your bend against the drawing, and continue bending and checking until they match. Repeat for the opposite end of the braces. Then, repeat for the other set of braces.

continued

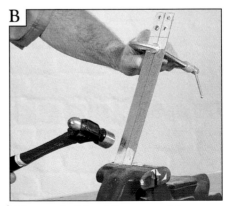

B

Align marked bend line on masking tape with top edge of vise jaws, and bend steel strap to correct angle.

The Fun-Time Racer

continued

4 Remove the tape to separate the pieces. Trace around a dime to mark the radius on each end of each brace. Grind the radii to shape and file or sand to remove the burrs.

5 Repeat steps 1 through 4 above to form the handle brackets, tongue brackets, upper and lower steering braces, and axle support retainers. See the Braces and Brackets Parts View drawing for shape and hole sizes.

ASSEMBLE THE REAR-AXLE AND STEERING ASSEMBLIES

1 Using a hacksaw, crosscut two pieces of ⅜"-diameter steel rod to 16" long for the axles.

2 Assemble and glue the rear axle assembly (C, D) in the configuration shown on the Rear Axle Assembly and Exploded View drawings. Repeat for the Steering Assembly and Handle. Check that the parts fit correctly.

ADD THE FINISH, AND LET THE GOOD TIMES ROLL

1 Remove the metal parts from the wood parts. Seal the plywood wheels and oak pieces with a clear exterior finish (we used Minwax exterior polyurethane).

2 Sand the metal parts to remove any burrs or sharp edges. Buff the surfaces of the metal parts with 00 steel wool to rough it up, and then wipe the parts clean with acetone to remove any residue.

3 Apply a coat of primer and two coats of satin black exterior paint to all the metal strapping.

4 Reassemble the metal parts to the finished wagon. Slide a ⅜" (I.D.) by 1" bronze bearing onto each axle where shown on the Wheel Assembly drawing.

5 Apply a bead of epoxy to the wheel coves, and stretch the tires onto the wheels.

6 Slip the wheels and washers onto the axles in the configuration shown on the Exploded View drawing. Secure the axles with ⅜" friction caps. For an added bit of flare, apply a decal to each side of the wagon. See the Buying Guide for our source.

BUYING GUIDE

• **Wagon hardware kit.** All the parts stated in the Supplies listing *except* for the dowel stock and finish. Kit no. WM-HW. For current price contact Miller Hardware 1300 M. L. King Parkway, Des Moines, IA 50314. Or call 515/283-1724 to order. No CODs please.

• **Wagon lumber kit.** All the individual oak and plywood pieces shown in the Cutting Diagram cut slightly oversized (length and width) from the thickness stated in the Bill of Materials. Stock no. W70. For current price contact Heritage Building Specialties, 205 North Cascade, Fergus Falls, MN 56537. Call 800-524-4184 to order.

• **Self-adhesive vinyl decals.** Two WOOD Racer decals with stripes. $17.95 ppd. Kness Signs, 5291 NW 72 Street, Urbandale, IA 50322. No phone orders please.

DISC SANDER SIDEKICK

Sanding rough-cut discs perfectly round will be easy once you build this sanding jig. We made ours in about 30 minutes.

To put your jig to work, clamp the base of the jig to your sander table, and center the disc to be sanded on the dowel near the end of the adjustment arm. Start the sander and slowly slide the adjustment arm with attached workpiece into the sanding disc until the disc sands to the marked circumference line on the workpiece. Turn the sander off, clamp a ¾ x ¾ x 2" stop to the adjustment arm, turn the sander on again, and slowly rotate the workpiece against the sanding disc.

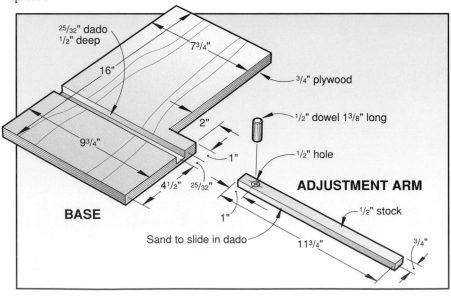

Small Gifts

If you're looking for the perfect gift, look no further. Whether it's a birthday, holiday, or you're just in the mood to surprise someone special, the small gifts in this chapter are appropriate for many occasions.

Cupid's Mystery Valentine

Carvers, you'll love this one

When you get right to the heart of the matter, this project comes down to one question: "How do you get that arrow through there?" The answer doesn't involve cutting or gluing, nor is there complicated carving involved. Curious? Read on.

Note: *Start with a ³⁄₄ × 2¹⁄₂ × 2³⁄₄" block for the heart and a ³⁄₈ × ³⁄₄ × 7¹⁄₄" piece of another species for the arrow. This proves you didn't carve the puzzler from a single block. (We carved a cedar heart and a basswood arrow.) Select quartersawn stock for the arrow; that is, stock with the end grain running straight across, as shown in the illustration* opposite page.

If you have trouble finding quartersawn ³⁄₈"-thick stock for the arrow, make your own, as we did. Start with a piece of basswood (or other) that has the end grain running almost straight across, like that shown in the Cutting the Arrow Stock drawing.

Select stock sufficiently thick and wide that you can draw one or two ³⁄₈ × ³⁄₄" rectangles on the end, long sides perpendicular to the grain. Stock that's about 1" thick should work, depending on the angle of the grain. For safety and convenience, start with a workpiece at least 12" long.

Starting at one corner, draw a line across the end of the stock, parallel to the grain (line 1, shown in red on the drawing). Draw line 2 perpendicular to line 1, ³⁄₄" long. Then, draw lines 3 and 4 where shown. If there's room on the end, lay out another rectangle.

Tilt your bandsaw table so that line 1 lies parallel to the blade. Bevel-rip the workpiece at lines 1 and 3. Then, set the bandsaw table

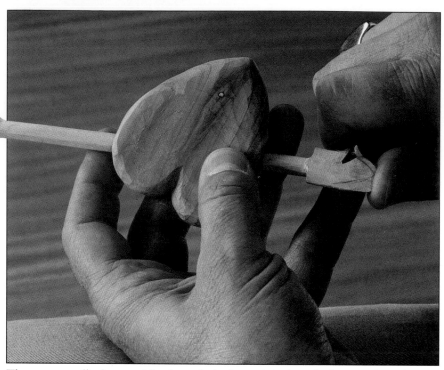

That arrow really does go right through the heart, it isn't trick photography. There is a bit of trick carving involved, though.

to 0° tilt (90° to the blade), and resaw the stock to ³⁄₈" thick, removing the wedge-shaped sides. Set the arrow stock aside for now.

COME ON, HAVE A HEART

Trace the Full-Sized Heart Front View pattern onto the carving stock for the heart. Bandsaw the blank. Mark the center for the ⁷⁄₁₆" hole where shown on the right edge of the blank.

Chuck a ⁷⁄₁₆" bit into your drill press. Then, align the diagonal centerline on the pattern with the drill bit, holding the blank at that angle with a handscrew clamp. Center the bit on the edge of the blank where shown, and drill.

Now, carve the heart. The carving is minimal—it amounts to little more than rounding over the edges. You can accomplish this in any number of ways, depending on the tools you have and the look you want to achieve—smooth or rough-

hewn. We rounded over the edges with a bench knife, and curved the front and back with gouges to make the heart ³⁄₈" thick at the point. Apply a clear finish.

THE SECRET OF THE ARROW

It's time to put the arrow through the hole. Figured it out yet? Here's how we did it, using a technique passed on to us by whittler Roald Tweet of Rock Island, Illinois.

Trace the Full-Sized Arrow pattern onto the ³⁄₈" stock. Be sure to place the arrowhead at the end that has grain running straight across. (We found on some of our stock that the grain didn't go straight across at both ends.)

Scrollsaw the blank, cutting on the *red* pattern line. Whittle the arrow shaft to about ¹⁄₄" diameter. Don't carve the ends just yet, though. Instead, place the rectangular end—the one that will be the arrowhead—in your vise, patterned face up.

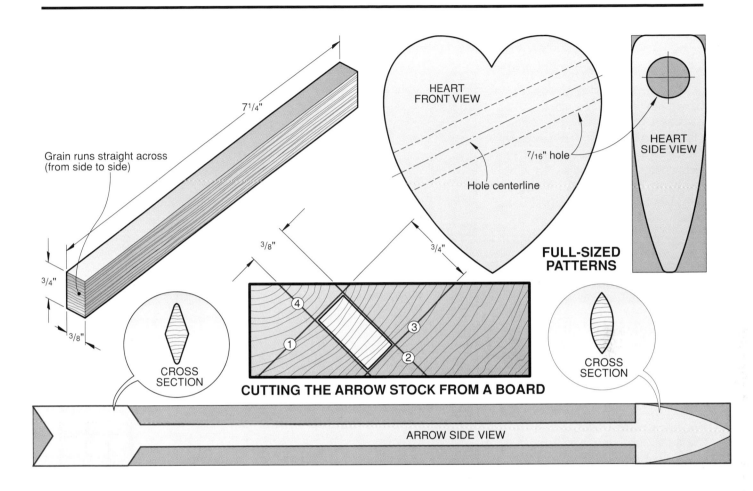

7 1/4"

Grain runs straight across
(from side to side)

3/4"

3/8"

CROSS
SECTION

HEART
FRONT VIEW

7/16" hole

Hole centerline

HEART
SIDE VIEW

3/8"

3/4"

① ② ③ ④

CUTTING THE ARROW STOCK FROM A BOARD

FULL-SIZED
PATTERNS

CROSS
SECTION

ARROW SIDE VIEW

Now, close the vise, crushing the wood as shown *right*. Squeeze the end of the arrow until it's narrow enough to slip through the hole in the heart. Whittle the corners off the compressed end to make it fit, as shown *far right*.

The wood won't always cooperate. It may not squeeze down enough, or it might even crack. You'll just have to start over with a new blank if things go awry.

Once you've placed the arrow through the hole, soak the crushed end in a glass of water. Soon, it will start to swell back to its original size. Depending on the wood, the end should be back to original size within an hour or so.

Let it dry. Then, carve the ends of the arrow, and finish.

Compress the arrowhead slowly and steadily, reducing the width to nearly the diameter of the arrow shaft.

Whittle away the corners of the compressed end until it will just fit through the hole in the heart.

Restaurant in the Round

Turners and birds both will love it

Expect a bevy of birds to gather 'round this clever feeder, which combines a few simple turnings with a section cut from a plastic pop bottle. Use rough cedar for a rustic look, or build one from redwood.

From ¾"-thick cedar or redwood stock, bandsaw or scrollsaw six discs, one 11" in diameter, two 9", one 7", one 6", and one 3". (We cut our cedar circles from a 1×12×48" board, even allowing for waste around knots.)

On the 7" disc, draw a line ½" in from the edge and parallel to it, creating a 6" circle and a ½"-wide ring around the outside, shown in the Cutting the Ring drawing. Cutting parallel to the grain, bandsaw in from the edge to separate the disc and the ring. Or, drill a blade start hole, and scrollsaw it.

Glue closed the entry cut through the ring. (We used cyanoacrylate adhesive. Woodworker's glue or epoxy would work also, but you'll need to clamp the ring with a band-type clamp or large rubber band.) Drill a ⅜" hole through the center of each disc.

CHOP A POP BOTTLE TO MAKE THE CYLINDER

Wash out a 2-liter plastic pop bottle—either a clear or tinted one will work fine. With a utility knife, slice the straight-sided center section from the bottle, cutting where shown in the Plastic Seed Holder illustration.

Drill or punch a series of ⅛" holes around each end where shown in the illustration. Then, wrap a piece of ¾"-wide masking tape around one end of the plastic cylinder, flush with the edge.

Following the full-sized pattern *opposite*, draw two seed openings on the tape on opposite sides of the cylinder. Cut them out with scissors or snips, and remove the tape. Set the feeder cylinder aside now while you turn the wooden parts for the top and bottom.

MAKE A FACEPLATE FOR THE DISCS

Attach a ¾"-thick, 8"-diameter scrapwood auxiliary faceplate to a metal faceplate 3–6" in diameter. Round down the wooden disc's edge, and turn the face true.

With the assembly still mounted on the lathe, bore a ⅜"-diameter hole through the center of the wooden faceplate. Start by pushing a small gouge into the center of the rotating workpiece (we used a ⁵⁄₁₆"

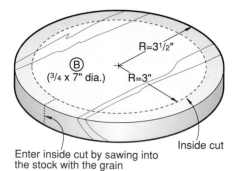

Enter inside cut by sawing into the stock with the grain

Inside cut

CUTTING THE RING

miniature turning gouge—other sizes also would work). Enlarge the hole to fit a $\frac{3}{8} \times 2\frac{1}{2}$" hex-head bolt.

Scribe or draw index marks on the lathe faceplate and the auxiliary faceplate, and remove the wooden one. Counterbore the backside of the wooden faceplate to accept the head of the bolt. Insert the bolt through the hole from the back of the auxiliary faceplate, and epoxy it in place, keeping the threaded end perpendicular to the front surface. After the glue cures, align the marks, and reattach the wooden faceplate to the metal one. Remount the assembly on the lathe.

TURN THE BIG DISC FIRST

Slide the 11" disc onto the bolt extending from the auxiliary faceplate. The top of the disc goes against the faceplate. Because the bolt threads don't reach the bottom face, slide the 3" disc onto the bolt as a packing piece. Secure with a hex nut or wingnut.

With a gouge, true the edge of the disc. The final diameter isn't critical, but try to keep it in the vicinity of $10\frac{1}{2}$–11". True the edge of the 3" disc, too. Taper it slightly, like a cork stopper.

Measure the diameter of the plastic cylinder (ours measured $4\frac{1}{4}$"). With the parting tool, cut a $\frac{1}{8}$"-wide groove about $\frac{1}{4}$" into the face of the disc to fit the cylinder.

Now, run the lathe at a medium speed (about 1000 rpm). Hold the cylinder perpendicular to the disc with the top toward it. (The top is the end without seed openings.) Briefly and lightly press the end of the cylinder into the groove. Heat generated by friction will melt the plastic, forming a rolled edge.

Remove the disc from the faceplate, and turn it over, placing the grooved side against the faceplate. Instead of using the nut to secure it this time, drive a pair of #6 x 1¼" drywall screws into the disc

continued

Two-liter plastic pop bottle

Cut middle out of bottle and discard top and bottom pieces

Drill $\frac{1}{8}$" holes $\frac{1}{4}$" apart, $\frac{1}{8}$" from edge on top and bottom of middle section

Bottle

SEED OPENING
FULL-SIZED PATTERN

Cut two openings for seed (see pattern)

PLASTIC SEED HOLDER

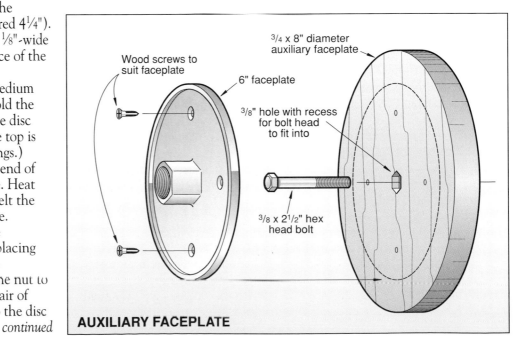

Wood screws to suit faceplate

6" faceplate

$\frac{3}{4}$ x 8" diameter auxiliary faceplate

$\frac{3}{8}$" hole with recess for bolt head to fit into

$\frac{3}{8}$ x 2$\frac{1}{2}$" hex head bolt

AUXILIARY FACEPLATE

Restaurant in the Round

continued

through the back of the faceplate. (We were able to do this without dismounting the faceplate.)

Slide the 3" disc onto the centerbolt with the largest diameter against the larger disc. Draw a pencil line around the small disc, and slide it off.

With the parting tool, cut about ⅛" deep straight into the wood inside the pencil line. Then, with a gouge, cut a hole through the disc to fit the 3"-diameter plug. Don't try for a tight fit; allow about ¹⁄₁₆" clearance all around.

Now, turn the edge to the cove profile shown on the *opposite page*. A ½" gouge will do the job. Be sure to form the smallest diameter on the top of the disc—the side facing out. Remove the disc from the faceplate.

GETTING TO THE BOTTOM OF IT

Slide a 9" disc then a 6" one onto the faceplate, the best side of each facing out. Align the grain on both, and then fasten them together with three #6 x 1¼" drywall screws driven in through the 6" disc. Place the screws inside an imaginary 2½"-diameter circle around the center.

Secure the discs to the faceplate with a nut or wingnut. True the edges, and turn the cove edge profile on each piece.

Remove the assembly from the faceplate, and replace with the small disc toward the inside. Secure the turning, and turn a groove for the bottle section as you did on the 11" disc. Roll the bottom edge of the plastic cylinder as you did previously.

Next, center the 7" ring around the groove. Glue it into position with waterproof glue. Turn the ring to the profile shown *opposite page*. Be careful as you work on the inside of the ring.

TURN THE LID FOR THE FEEDER

Slide the remaining 9" and 6" discs onto the faceplate as you did previously. Screw them together, and turn the edge profiles as before.

Turn the assembly around on the faceplate, and center the 3" plug on the underside by sliding it on over the bolt. Glue the plug in place with waterproof glue.

TWO FINIALS FINISH IT OFF

Remove the faceplate from the lathe, and install a spur-type drive center and a tail center to turn the finials. A cone-type revolving center will work best in the tailstock.

Locate the center on each end of your 2½ x 2½ x 12" stock. (For contrast with the discs, we turned our finials from fir, which we sawed from a piece of 4 x 4. You also could laminate material for the finials.) Mount the stock and round it down to 2⅜" diameter.

Lay out two finials on the cylinder, using the full-sized template on the *opposite page*. Here's one way to do it: Place them base-to-base, turning the top of the first one 2" from the headstock end of the cylinder. Leave about ¼" between the bottoms of the bases.

Turn the sphere with a gouge or 1" skew, working from the largest diameter down. The bullseye pattern formed by the woodgrain will aid you in turning a true sphere rather than an egg. Just watch for a round target.

After shaping both finials, turn a small-diameter tenon (about ¼" works fine) at the top of each sphere and between the bases. Remove the workpiece from the lathe, then cut off the waste and separate the turnings from each other with a backsaw.

Now, drill the ⅜" hole through the center of each finial. Grip the turning with a handscrew clamp to drill the hole with a drill press.

PUT IT ALL TOGETHER

With epoxy, glue the plastic cylinder into the grooved discs. The top, the end without the seed openings, fits into the groove on the largest disc. Fill the groove with epoxy, then insert the cylinder into it. The epoxy will flow through the holes around the rim to aid in making a strong joint. Glue on the bottom the same way. Glue one finial to the lid, the other to the bottom. Align the holes using the auxiliary faceplate.

Assemble the bird feeder on a ⅜"-diameter brass, aluminum, or galvanized steel rod 36" long. First, drill a ⁵⁄₃₂" hole ¼" from each end of the rod. Slide a 1"-diameter splitring key ring into one hole. Place the bird feeder body on the rod, finial pointing down, and slide it down. Place the lid on the rod, and install the other ring in the hole at the top of the rod. Hang the feeder by the top ring. To fill it up, slide the lid up the rod and pour seed in through the hole in the top of the feeder.

TOOLS AND SUPPLIES

- **Stock**
  1 x 12 x 48" (nominal dimensions) cedar or redwood, 2½ x 2½ x 12" pine or fir
- **Lathe equipment**
  Spur drive center, rotating cone tail center, 3–6" faceplate
- **Tools**
  ⅛" parting tool; ⁵⁄₁₆", ⅜", and ½" gouges; ½" and 1" skews
- **Lathe speeds**
  When truing the discs, particularly the large one, run your lathe at 500–800 rpm. Increase the speed to 1200–1500 rpm for the profile turning.

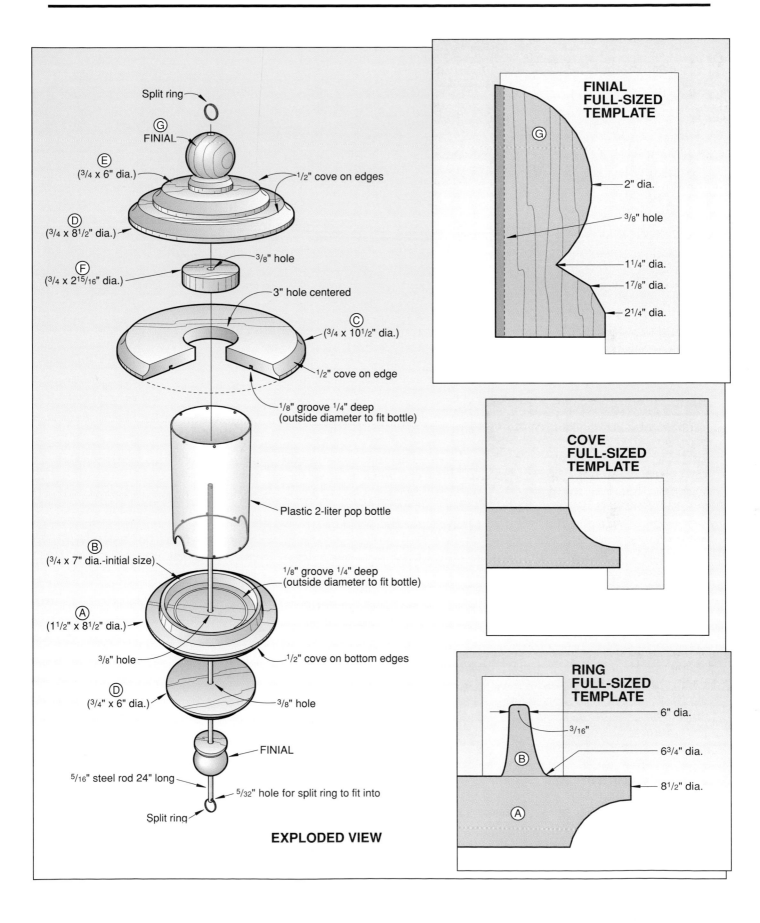

Split ring

G
FINIAL

E
(³/₄ x 6" dia.)

¹/₂" cove on edges

D
(³/₄ x 8¹/₂" dia.)

F
(³/₄ x 2¹⁵/₁₆" dia.)

³/₈" hole

3" hole centered

C
(³/₄ x 10¹/₂" dia.)

¹/₂" cove on edge

¹/₈" groove ¹/₄" deep
(outside diameter to fit bottle)

Plastic 2-liter pop bottle

B
(³/₄ x 7" dia.-initial size)

¹/₈" groove ¹/₄" deep
(outside diameter to fit bottle)

A
(1¹/₂" x 8¹/₂" dia.)

³/₈" hole

¹/₂" cove on bottom edges

D
(³/₄" x 6" dia.)

³/₈" hole

FINIAL

⁵/₁₆" steel rod 24" long

⁵/₃₂" hole for split ring to fit into

Split ring

EXPLODED VIEW

**FINIAL
FULL-SIZED
TEMPLATE**

G

2" dia.

³/₈" hole

1¹/₄" dia.

1⁷/₈" dia.

2¹/₄" dia.

**COVE
FULL-SIZED
TEMPLATE**

**RING
FULL-SIZED
TEMPLATE**

6" dia.

³/₁₆"

B

6³/₄" dia.

8¹/₂" dia.

A

Turned Perfume Applicator

An easy-to-make gift that looks, feels, and smells great

Maybe you got away with giving a steam iron last year, but don't push your luck this time. Instead, show your sensitive side with this purse-sized perfume applicator. And, because you can turn it yourself, it's a gift with extra meaning.

Start with two pieces of ¾"-square turning stock, one 2¹⁵⁄₁₆" long and the other 1". The longer part will be the perfume applicator's body; the shorter one, the cap. For a great-looking turning with grain that appears to be continous from end to end, cut both parts from a single piece of stock. Then, mark the mating ends and keep the pieces in order for the next steps.

Locate the center on one end of each piece. Chuck a ⁵⁄₁₆" bit in your drill press, and drill lengthwise through each piece. Grip the stock with a handscrew clamp, and keep it parallel to the bit.

Glue a brass tube (supplied in the kit) into each drilled blank, using either epoxy or cyanoacrylate (instant) adhesive. Don't starve the ends of the tube for glue—the brass and wood must be bonded firmly at the ends. Sand the wood flush with the brass on the ends, keeping the ends square.

TURN THE CAP AND BODY

Slide the blanks onto the mandrel, separating them with the steel bushings where shown in the Turning the Cap and Body to Shape drawing, *opposite page*. Thread on the nut, and tighten it sufficiently to keep the wood from turning on the shaft.

Now, mount the mandrel between centers on your lathe. Grip the unthreaded end in a Jacobs chuck or a three-jaw chuck on the headstock. Insert the point of a rotating tail center into the centerbore on the other end.

Round down the blanks to ⅝" diameter with a ¾" or ½" gouge. Lay out one of the full-sized profiles shown *below* (or a design of your own) on the cylinders. Turn the body and cap profiles. We used ¼" and ½" gouges and a ½" skew. At each end of each piece, turn the wood to the bushing diameter—⅜"—to match the fittings. Take shallow cuts as you approach ⅜" outside diameter; the wood will be only about ¹⁄₃₂" thick.

PERFUME APPLICATORS
Full-Sized Patterns

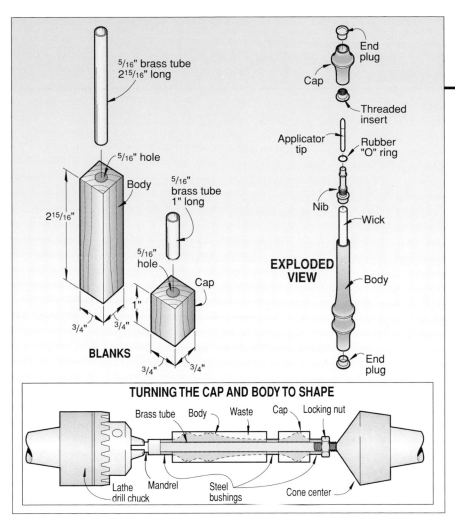

BLANKS

5/16" brass tube 2 15/16" long

5/16" hole

Body

2 15/16"

3/4" × 3/4"

5/16" brass tube 1" long

5/16" hole

Cap

1"

3/4" × 3/4"

EXPLODED VIEW

End plug
Cap
Threaded insert
Applicator tip
Rubber "O" ring
Nib
Wick
Body
End plug

TURNING THE CAP AND BODY TO SHAPE

Brass tube Body Waste Cap Locking nut

Lathe drill chuck Mandrel Steel bushings Cone center

Press fittings into tubes with a vise. Wood block shown here prevents damage to applicator nib.

With the lathe turned off, sand in the direction of the grain with progressively finer grits from 150 to 320. Apply a clear finish to the mounted parts. Note the orientation of the body and cap turnings, then remove them from the mandrel when dry.

ASSEMBLE THE APPLICATOR
Drill a 5/16" hole through a piece of 3/4"-thick scrapwood about 2×3". With double-faced tape, affix the block to the back jaw of your vise.

Refer to the Exploded View drawing, and assemble the applicator body. With your thumb, press the end plug into the end of the body tube that was at the headstock end of the mandrel.

Insert the large wick and nib into the other end. Place the extended tip of the nib into the hole in the block on the vise jaw as shown *above right*. Seat the nib's shoulder on the block, then slowly close the vise to press the nib and end plug into place.

Install the O-ring where shown, and tightly screw the threaded insert for the cap onto the nib. Place an end plug into the end of the cap tube that was nearest the tailstock. Now, fit the other end over the threaded insert screwed onto the nib. Align the cap grain with the body, then push the cap onto the insert. Carefully unscrew the assembly, and press the parts home with the vise.

Slip the applicator tip into the nib. (It fits loosely at first, but swells when you fill the applicator.) To fill, dip the tip into the fragrance for about a minute. A filling lasts for about a month.

BUYING GUIDE
• **Parts kits, mandrel.** Lathe mandrel, applicator kits, and 5/16" parabolic-flute drill bit (best bit for end grain) available from Berea Hardwoods Co., 6367 Eastland Rd., Brookpark, Ohio 44142. Or call 216/234-7949.

TOOLS AND SUPPLIES
• **Stock**
Miscellaneous hardwoods, 3/4 × 3/4 × 2 15/16" and 3/4 × 3/4 × 1" (Maple and purpleheart shown in photo)
• **Lathe equipment**
Three-jaw or Jacobs-type chuck for headstock, rotating cone center for tailstock, turning mandrel (see the Buying Guide)
• **Turning tools**
Spindle gouges, 1/4", 1/2", and 3/4" Skew, 1/2"
• **Lathe speeds**
Roughing, turning: 1000–1500 rpm

A Box That's a Bit Fishy

Scrollsaw this curious container

It takes just a few simple scrollsaw cuts to create this entertaining box. But, there's a catch—you must make those cuts in a specific order. Don't worry, though; we've mapped out all the moves for you, so you can cut out this puzzler in short order.

Note: We cut our box from 1¾" square stock. If your scrollsaw won't accept material that thick, you can adapt the design to fit smaller square stock. For cutting, we used a no. 5 blade, .037×.015" with 14 teeth per inch.

Referring to the full-sized Side View pattern, *opposite page*, draw the red outside cutting line, designated no. 1, on one face of a 1¾×1¾×6" piece of stock. (We made one box of walnut and another of Philippine mahogany.) Saw around the line. Sand the edges as necessary.

Stand the workpiece on the bottom—the flattest edge, and mark a cutting line along the curved upper surface ¼" from the back, shown by the *green* line (no. 2) on the Top View drawing. Saw along the line to cut off the thin back piece. Set the back aside to use later.

Now, draw the removable keys, the *yellow* lines shown as no. 3 on the Side View pattern. Scrollsaw the keys—the mouth and the tail. Feed the work into the saw carefully and slowly to avoid flexing the blade excessively. Bending the blade as you cut results in tapered pieces that may not slide apart readily. Set the keys aside, and turn the workpiece onto the flat bottom again.

Draw the *purple* cutting line for the lid (no. 4) onto the curved surface facing up. Scrollsaw along that line, then slide the lid piece off. Set it aside.

With the lid off, draw the *blue* interior cutting line (no. 5) where shown. Drill a ¹⁄₁₆" blade start hole where shown, then insert the scrollsaw blade and cut out the center section of the body.

On the cutout part, draw a line about ³⁄₁₆" below the keyed edge and parallel to it, shown by line no. 6 on the Inner Lid and Supports drawing. Cut off the inner lid, then saw the ends from the remaining piece where shown by lines no. 7.

To assemble the box, first glue the thin back piece to the body with woodworker's glue. Align the front end and edges, and clamp. Remove any squeeze-out in the key slots or the interior.

Glue the inner lid supports into position at the front and back of the interior cavity. You can use woodworker's glue or cyanoacrylate adhesive for these parts.

After the glue dries, sand as necessary. Set the inner lid into place, slide the lid onto the body, and drop the keys into position. Sand the outside of the box. Remove the keys and lids, then apply a clear oil finish on all parts, taking care not to build it up so thick that the parts won't fit together. When dry, reassemble the box.

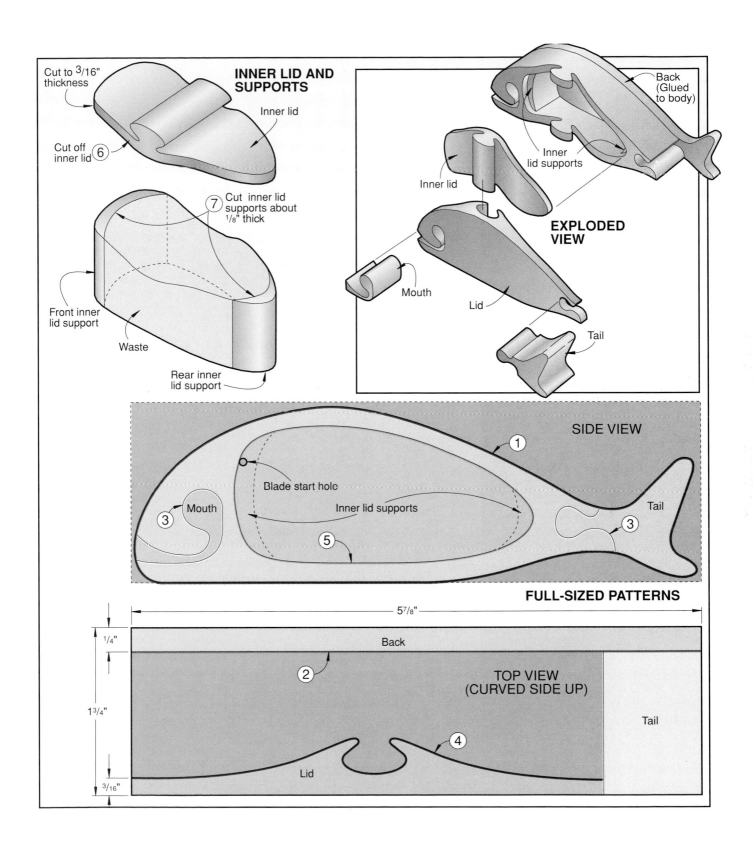

INNER LID AND SUPPORTS

Cut to ³/₁₆"
thickness

Inner lid

Cut off
inner lid ⑥

⑦ Cut inner lid
supports about
¹/₈" thick

Front inner
lid support

Waste

Rear inner
lid support

Back
(Glued
to body)

Inner
lid supports

Inner lid

**EXPLODED
VIEW**

Mouth

Lid

Tail

SIDE VIEW

①

Blade start hole

Inner lid supports

③ Mouth

⑤

Tail

③

FULL-SIZED PATTERNS

5⁷/₈"

¹/₄"

Back

②

**TOP VIEW
(CURVED SIDE UP)**

1³/₄"

Tail

④

³/₁₆"

Lid

Block Prints

Carvings you can share with your friends

PRINTING AND CARVING HAVE BEEN TOGETHER FOR CENTURIES

Back about 770 A.D., historians say, Empress Shotoku of Japan ordered a million prayer cards. For a job that big, hand-copying was out. So, an artisan carved mirror-image, raised characters on a wooden block. He spread ink on the raised surface, pressed a piece of paper against it, and thus invented printing.

About 950 A.D., paper money was being printed from wooden blocks in the Szechwan province of China. Wood-block printing appeared in Europe around 1300. Later, movable metal type pushed carved wood to

the background in the printing trade. Wood came back with the rise of illustrated magazines in the 19th century.

In those popular periodicals, detailed drawings hand-engraved on wood depicted great feats or cataclysmic events. With the advent of photography and photoengraving, though, woodcut illustrations faded from the commercial printing scene. Artists have kept the process alive, and many make block prints today.

As much as you'd like to, you **can't send a carving to everyone on your Christmas list. So, here's the next best thing:**

Relief-carve a seasonal design and print your own cards from it. You don't have to be an expert carver—or printer—to achieve great results with this old printmaking technique.

For fine-arts printmakers, wood has long been a favorite medium. Most artists working in wood-block printing make *woodcuts*, designs carved on the face of a piece of wood with knives and gouges or power-carving equipment. *Wood engraving*, another style, involves incising the image into the end grain of a block with engraver's tools.

FULL-SIZED PATTERN
Carve away dark shaded areas

Woodcuts generally feature bold areas with few fine lines. Wood engravings are more detailed, often looking like shaded Victorian magazine illustrations. For our Christmas cards, we chose the woodcut style.

PRINT WITH PLYWOOD

The process starts with a printing block, which is nothing more than a relief carving. And, it can be a shallow relief at that. For the cards shown, we carved less than 1/16" deep on 1/4"-thick plywood.

The thickness of the block isn't critical, but it does need to be flat and have one smooth face without splits or knots. With judicious cutting, scraps of ordinary A-B or A-C plywood will work fine. You could use solid stock, too.

Cut your blank to 3¼ × 6¼". Saw it slightly oversize, then sand to finished size. This way you'll avoid splintered edges or ends, which could mar your prints. Sand the good face smooth. This will be your printing surface, so you don't want any dings or dents to spoil your prints.

GIVE THAT PATTERN A FLIP

Whatever you carve into the block will come out printed *backwards* on your paper. So, you must reverse the pattern as you transfer it to the wood. (Direction doesn't matter much with the patterns we show, but it could on others.) Here's how.

Photocopy one of the three patterns *above and next two pages*, with the machine set for a dark copy. Lay the copy facedown on your printing block, and run a hot clothes iron (the wool setting works fine) over the back to transfer the image to the wood.

Or, lay onionskin (tracing paper) over the original pattern. Trace the pattern lines, using a fine-point marker. Now, flip the tracing over, placing the side you drew on against the block. Fasten it with tabs of masking tape at the top. Then, slip a piece of transfer paper or carbon paper between the wood and the tracing, and trace the lines onto the wood through the thin paper.

The pattern you carve not only will print reversed, it will print as a *negative image*. That is, the dark lines and areas on the pattern illustration will be white (or whatever your paper color) in the print. Conversely, the light areas on the drawing will print in your ink color.

CARVE YOUR CHRISTMAS CARD

Cut away the lines and dark areas in the pattern. We carved our blocks with palm-handled tools and a bench knife. Use whatever tools you prefer, but be sure they're sharp. Ragged cuts in the carving translate to fuzzy lines in the finished print. Hold the block with a bench hook or nonskid mat as you carve.

Begin with the inside detail lines, such as the angel's face. Cut the narrow lines with a V-tool. Maintain constant width and depth for uniform lines.

For larger areas, such as the trees or the backgrounds, first stop-cut along the pattern line with a knife or V-tool. A *stop-cut* is simply a vertical incision along the pattern line. It breaks the chip as you cut toward the line.

continued

Block Prints

continued

1. Apply a thin, uniform coat of ink to the block with a brayer. Spread it across the raised surface, but don't get ink in the lines or carved areas.

2. Rub the back of the print evenly for the best impression. A bamboo baren, left in photo, produces the best print, but a spoon bowl does a passable job.

3. Peel the print back slowly from a corner to remove it from the block. Inspect it as you go. Be careful not to smear the ink at this stage.

After making your stop cut, start in the dark area and work toward the pattern line with a shallow gouge, such as a no. 5. Clear out all waste in the area, working from the center out to all edges. Cut only as deep as your stop-cut; otherwise, you'll probably chip out the edge of the carving area. Where gouges won't work, use the tip of your knife. Remember, the relief doesn't need to be deep, and the background doesn't need to be smooth.

PAPER TO PRINT ON

"For block printing, Japanese rice paper is best," according to Des Moines, Iowa, print artist Peggy Johnston, who designed our woodcuts. "It isn't stiff enough to fold into a card, though," she hastily adds. "A lot of artists print on a piece of rice paper, then mount it on a stiffer piece."

Peggy printed on $3\frac{1}{4} \times 6\frac{1}{4}$" pieces cut from a 19×24" sheet of white rice paper (three across, seven vertically). Then, with spray

adhesive, she mounted each print on the front flap of a $6\frac{1}{2} \times 7$" piece of drawing paper folded in half the long way. The finished cards fit regular $3\frac{5}{8} \times 6\frac{1}{2}$" envelopes sold by drugstores and stationers.

You also can print directly onto the heavier folded paper, as shown in the photographs. (For two-color cards, print onto colored stock.) If you print right onto the folded card, open it up before you print so you're working with only one thickness.

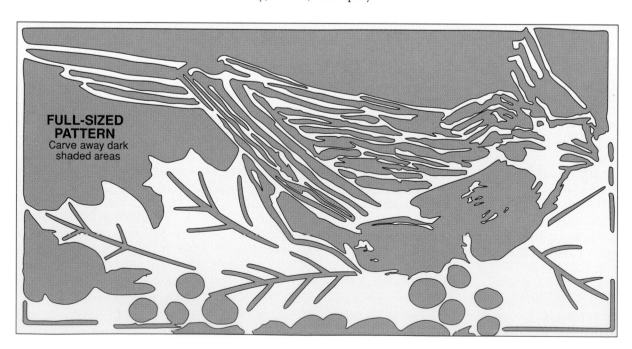

FULL-SIZED PATTERN
Carve away dark shaded areas

NOW, IT'S TIME FOR A PROOF

With the block carved and a supply of paper on hand, you're ready to print. Start with a *proof,* which is just another name for a test print. You'll check the proofs for any needed corrections.

First, squeeze out a bead of block-printing ink (1" or so) onto the center of a piece of glass or acrylic sheet. We used water-soluble ink for easier clean-up. Then with a *brayer,* a rubber roller with a short handle, roll out the ink into a thin film.

Now, ink the printing block with the brayer. Roll it across the spread-out ink to pick some up, then over the carved surface of the printing block to deposit some, as shown in Photo 1, *opposite page.* Roll from several directions, covering the raised area completely. Don't get ink in the lines or background.

Align an edge of the paper along one edge of the block without letting it touch the inked surface. Then, lay the paper facedown across the block. Don't try to move it after you've laid it down. Rub the back with a bamboo *baren* (a printmaker's rubbing pad) or the back of a spoon, as shown in Photo 2, *opposite page.* Hold the paper to prevent it from shifting as you rub firmly, covering the entire image area.

Lift one corner, as shown in Photo 3, *opposite page.* Peel the print back slowly to make sure you've printed the entire design. After you've removed the print, check it closely. If you need to touch up the carving anywhere, wipe the ink off the block first. Print another proof after making any changes. Continue proofing and correcting until you're satisfied, then print the cards.

String a clothesline to hang up the completed prints to dry. Ink the block, and produce your first print. Hang it up to dry (we used regular clothespins, hanging two prints back-to-back with each one). Then, re-ink the block and print another card.

For a numbered edition, keep the prints in order as you lift them off. After the ink dries, count how many you've made—say 42. Number the first one 1/42, the second 2/42, and so on. Pencil the number in the margin, along with your signature and the date.

FULL-SIZED PATTERN
Carve away dark shaded areas

TOOLS AND SUPPLIES

- **Stock**
  $1/4 \times 3\frac{1}{4} \times 6\frac{1}{4}$" plywood, one smooth face
- **Carving tools**
  V-tool:
    $\frac{1}{16}$", $\frac{1}{8}$" no. 12
  Gouges:
    No. 11 U-veiner, $\frac{1}{8}$"
    No. 9, $\frac{1}{4}$"
    No. 5, $\frac{1}{8}$", $\frac{1}{4}$"
  Knife:
    Bench knife or X-Acto knife
- **Printing supplies**
  Block-printing ink, water-soluble, your choice of color
  Brayer
  Baren or tablespoon
  Glass or acrylic, about 12×12"
  Paper (see text)

BUYING GUIDE

- **Printing supplies and paper.** For ink, brayer, baren, rice paper, and drawing paper, see your local art-supply dealer, or order from The Art Store, 600 Martin Luther King Blvd., Des Moines, IA 50312, 800/652-2225.

Tabletop Reindeer

These festive accents will steal the show

You may start out cutting just a couple of these adorable little guys, but we'll bet you something. Before you're done, you will have geared up and made a whole herd for your friends and relatives. They make delightful and inexpensive little gifts and work great as seasonal centerpieces.

1 Using carbon paper or photocopies of the patterns and spray adhesive, transfer the full-sized body patterns on the *opposite page* to posterboard as shown in the cutting diagram. Cut the patterns to shape

to form templates. (Since we planned on making several reindeer, we made templates. If you're making just a set or two, transfer the full-sized patterns directly onto ⅛" stock.)

2 Use the templates to transfer the patterns to ⅛" stock. (See the Buying Guide for our source of void-free birch plywood.)

3 Cut the pieces to shape (we used a scrollsaw and a #2 blade— .029 × .012"—with 20 teeth per inch; you also could use a bandsaw equipped with a ⅛" blade). When cutting the notches in the parts, remember that the notch width needs to be the same as the thickness

of the stock you're using. (Since we used ⅛" material for our deer, we show ⅛"-wide notches on our full-sized patterns; adjust if necessary.)

4 Lightly sand the edges and surfaces smooth with 220-grit sandpaper. Assemble each deer (we used a drop of instant glue at each joint to hold the pieces together). Spray on several light coats of gloss enamel. (To avoid runs, we found it essential to apply light coats rather than try to paint the deer with a heavy coat or two.)

BUYING GUIDE
• **Void-free ⅛" birch plywood ⅛ x 12 x 15" (enough for two deer).** For current prices contact Heritage Building Specialties, 205 North Cascade, Fergus Falls, MN 56537. Call 800/524-4184 to order.

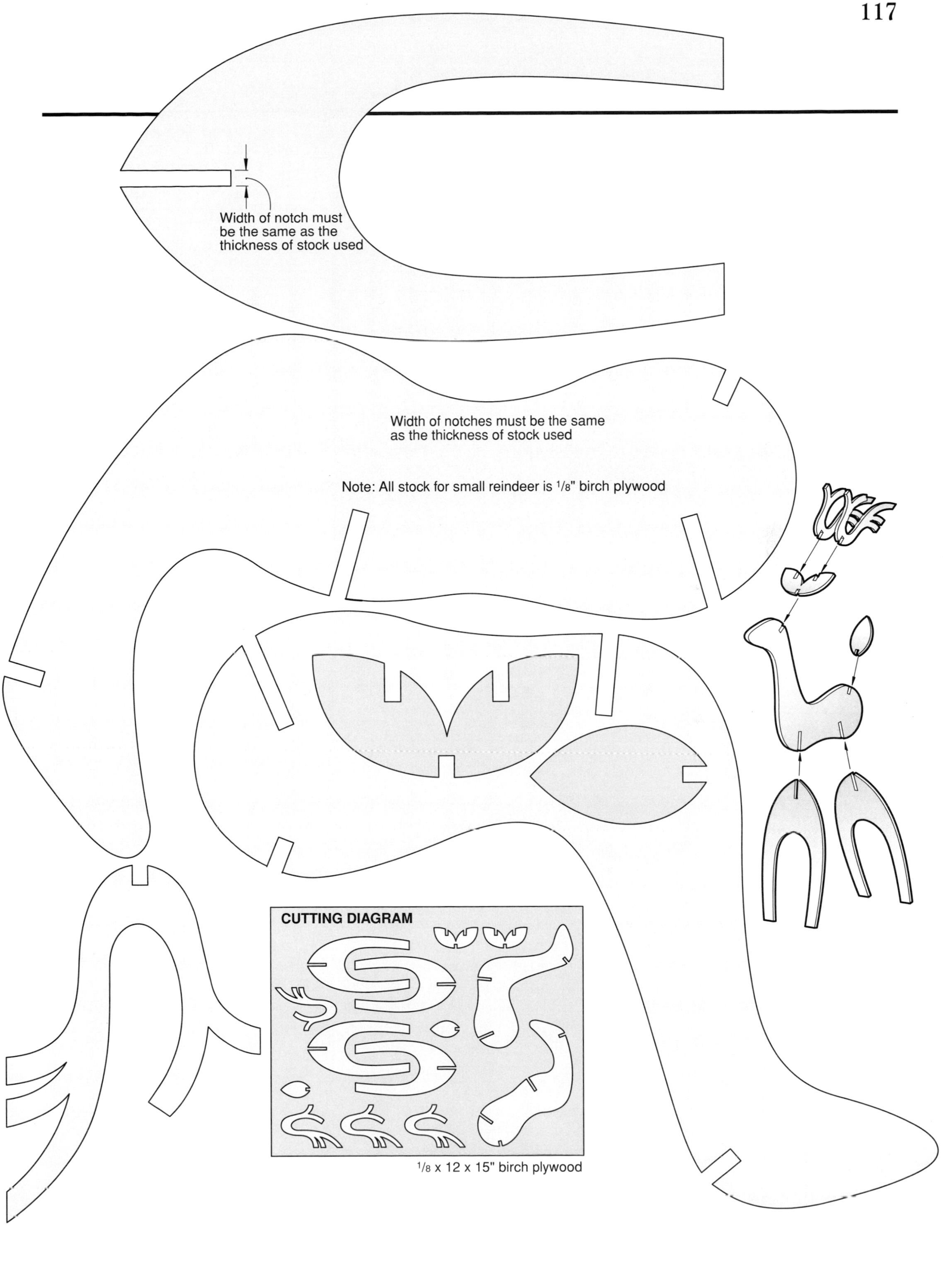

Width of notch must
be the same as the
thickness of stock used

Width of notches must be the same
as the thickness of stock used

Note: All stock for small reindeer is 1/8" birch plywood

CUTTING DIAGRAM

1/8 x 12 x 15" birch plywood

Dresser-Top Delight

Twelve roomy drawers offer ample storage for rings, necklaces, and watches

Bill of Materials

Part	Finished Size*			Mat.	Qty.
	T	W	L		
BASIC ASSEMBLY					
A* sides	3/8"	7 1/2"	13"	EC	2
B* top	1/2"	7 7/8"	13 1/2"	EC	1
C shelves	3/8"	7 3/8"	12 3/8"	EC	6
D bottom rail	3/8"	1 9/16"	12"	C	1
E back	1/8"	12 3/8"	12 13/16"	BP	1
TRIM					
F* front	1/2"	1 3/4"	13 3/4"	C	1
G* sides	1/2"	1 3/4"	8"	C	2
DRAWERS					
H* fronts	3/8"	1 7/16"	4"	C	6
I* fronts	3/8"	1 7/16"	6"	C	4
J* fronts	3/8"	1 7/16"	12"	C	2
K sides	3/8"	1 7/16"	7 5/16"	C	24
L backs	3/8"	1 3/32"	3 5/8"	C	6
M backs	3/8"	1 3/32"	5 5/8"	C	4
N backs	3/8"	1 3/32"	11 5/8"	C	2
O bottoms	1/8"	3 5/8"	6 1/8"	BP	6
P bottoms	1/8"	5 5/8"	6 1/8"	BP	4
Q bottoms	1/8"	11 5/8"	6 1/8"	BP	2
DRAWER GUIDES					
R top	1/4"	3 7/32"	7"	H	6
S middle	1/4"	5 7/32"	7"	H	4
T bottom	1/4"	11 7/32"	7"	H	2

*Initially cut parts marked with an * oversized. Trim to finished size according to the instructions.

Materials Key: EC–edge-joined cherry, C–cherry, BP–birch plywood, H–hardboard.

Supplies: #17x3/4" brads, #18x1/2" brads, clear finish.

Show off your craftsmanship with a jewelry showcase that features drawers with handsome splined joints. To help you create this accent, we've included the plan for a simple corner-kerfing jig and instructions showing you how to put it to work.

Note: *You'll need several thicknesses of solid stock for this project (we used cherry). You can plane or resaw thicker stock to the thicknesses listed in the Bill of Materials, or see the Buying Guide for our source of preplaned cherry or walnut stock.*

CUT THE SIDES AND TOP PIECES FIRST

1 Cut the side panels (A) to 7 5/8 × 13". Now, rip a strip 3/8" wide off the front edge of each side panel.

Crosscut the narrow strip to 12 13/16" long. See the Side Panel drawing for reference.

2 Lay out the locations, and cut 3/8" dadoes 3/16" deep on the inside face of each side panel where dimensioned on the Side Panel drawing. Then, cut a 3/16" rabbet 3/16" deep along the top outside edge of each piece. Next, cut a 3/16" rabbet 1/8" deep along the back inside edge of each side panel.

3 To hide the dadoes showing on the front edges of each side panel, glue the 3/8 × 3/8 × 12 13/16" strip (trimmed from the front edge in Step 1) against the front edge of each panel, keeping the surfaces flush and the top end of the strip flush with the shoulder of the rabbet. Later, remove the clamps, and sand the panels.

4 Cut the top (B) to 8 × 13 1/2". Rip 3/4" from the front edge where

shown on the Top drawing on the *opposite* page, and set the narrow piece aside for now.

5 Mark the locations and cut a pair of 3/16" dadoes 3/16" deep on the bottom side of the top piece. Glue the 1/2 × 3/4 × 13 1/2" strip against the front edge of the top piece. Later, remove the clamps and sand smooth.

6 Rout a 1/4" cove along the bottom front and side edges of the top piece. Wrap sandpaper around a 1/2" dowel and sand the coves smooth.

continued

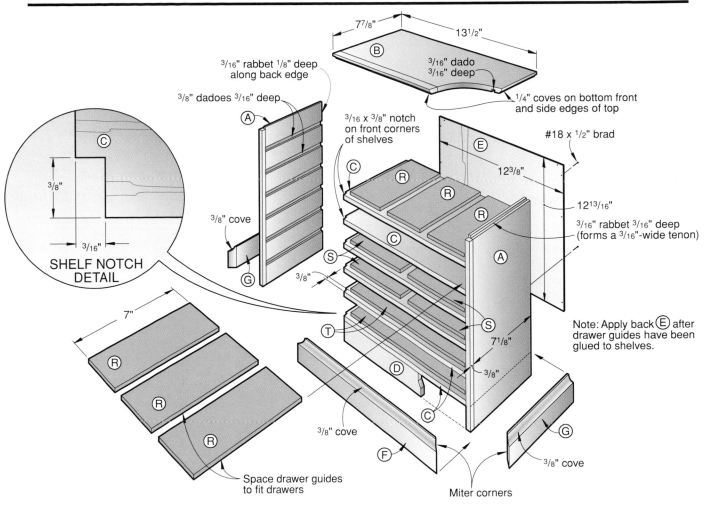

$^3/_{16}$" rabbet $^1/_8$" deep
along back edge

$^3/_8$" dadoes $^3/_{16}$" deep

$^3/_{16}$ x $^3/_8$" notch
on front corners
of shelves

$^3/_8$" cove

SHELF NOTCH DETAIL

$^3/_8$"

$^3/_{16}$"

7"

Space drawer guides
to fit drawers

$^3/_8$"

7$^7/_8$"

13$^1/_2$"

$^3/_{16}$" dado
$^3/_{16}$" deep

$^1/_4$" coves on bottom front
and side edges of top

#18 x $^1/_2$" brad

12$^3/_8$"

12$^{13}/_{16}$"

$^3/_{16}$" rabbet $^3/_{16}$" deep
(forms a $^3/_{16}$"-wide tenon)

Note: Apply back Ⓔ after
drawer guides have been
glued to shelves.

7$^1/_8$"

$^3/_8$"

$^3/_8$" cove

$^3/_8$" cove

Miter corners

EXPLODED VIEW

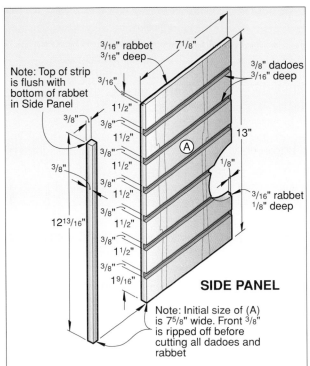

Note: Top of strip
is flush with
bottom of rabbet
in Side Panel

$^3/_{16}$" rabbet
$^3/_{16}$" deep

7$^1/_8$"

$^3/_8$" dadoes
$^3/_{16}$" deep

$^3/_{16}$"

$^3/_8$"

1$^1/_2$"

$^3/_8$"

1$^1/_2$"

$^3/_8$"

13"

1$^1/_2$"

$^3/_8$"

$^1/_8$"

$^3/_{16}$" rabbet
$^1/_8$" deep

1$^1/_2$"

$^3/_8$"

1$^1/_2$"

$^3/_8$"

12$^{13}/_{16}$"

1$^1/_2$"

$^3/_8$"

1$^9/_{16}$"

SIDE PANEL

Note: Initial size of (A)
is 7$^5/_8$" wide. Front $^3/_8$"
is ripped off before
cutting all dadoes and
rabbet

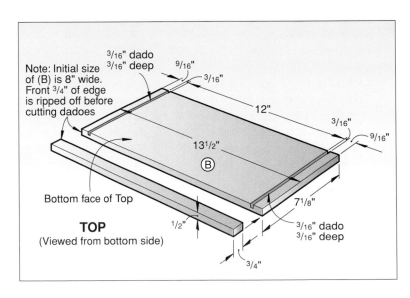

Note: Initial size
of (B) is 8" wide.
Front $^3/_4$" of edge
is ripped off before
cutting dadoes

$^3/_{16}$" dado
$^3/_{16}$" deep

$^9/_{16}$"

$^3/_{16}$"

12"

$^3/_{16}$"

$^9/_{16}$"

13$^1/_2$"

Bottom face of Top

TOP
(Viewed from bottom side)

$^1/_2$"

$^3/_4$"

7$^1/_8$"

$^3/_{16}$" dado
$^3/_{16}$" deep

Dresser-Top Delight

continued

ADD THE SHELVES AND BACK

1 Cut the six shelves (C) to size.

2 Mark and cut a 3/16 × 3/8" notch on the front corners of each shelf. See the Shelf Notch detail accompanying the Exploded View drawing for reference.

3 Dry-clamp the parts (A, B, C) to check the fit. Measure the opening and then cut the bottom rail (D) to size. Glue and clamp the parts, checking for square.

4 Measure the opening, and cut the back (E) to size from 1/8" birch plywood. Set the back aside for now; we'll attach it later.

LET'S CUT AND ATTACH THE TRIM NEXT

1 Cut a 1/2" piece of cherry to 1 3/4" wide by 32" long. Now, rout a 3/8" cove along one edge.

2 From the 32"-long piece, miter-cut the front trim piece (F) and the side trim pieces (G) to length.

3 Glue and clamp the trim pieces to the case. Wrap sandpaper around a 3/4" dowel and sand the coves smooth.

ADD A DOZEN DRAWERS FOR LOTS OF STORAGE

Note: *We constructed our drawers using the sizes of pieces listed in the Bill of Materials, creating a gap-free fit of the drawers in the case. Then, after positioning the drawer guides (R, S, T), we used a sanding block to sand the sides of each drawer for consistent 1/32" gaps where shown on the Front View drawing on page 123.*

1 From 3/8"-thick stock, rip 1 7/16"-wide strips for the drawer fronts (H, I, J). As shown on the Cutting Diagram, cut adjoining drawer fronts end-to-end from the same piece of stock. Doing this will allow side-by-side drawers to have continuous grain across their fronts.

CUTTING DIAGRAM

3/8 × 9 1/4 × 96" Cherry

(2) 1/8 × 3/8 × 24" for splines

3/8 × 9 1/4 × 96" Cherry

1/2 × 5 1/2 × 60" Cherry

See Buying Guide for our source of preplaned stock.

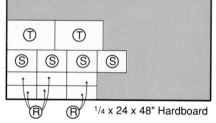

1/4 × 24 × 48" Hardboard

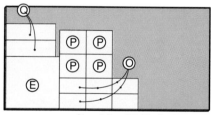

1/8 × 24 × 48" Birch plywood

2 Using the Drawer drawing for reference, cut a 1/8" groove 3/16" deep along the back edge of the long strips to be used for the drawer fronts.

3 Miter-cut the drawer fronts (H, I, J) to length.

4 Cut long lineal stock for the drawer sides (K). Cut a 1/8" groove 3/16" deep 7/32" from the bottom edge of the long drawer-side stock. Position the groove 7/32" from the bottom edge to create a 1/32" gap between the bottom edge of the drawer fronts and the top edge of the shelves. Cut and miter-cut the drawer sides to length. For housing the drawer backs later, cut a 3/8" dado in each drawer side where dimensioned on the drawing.

5 Cut the drawer backs (L, M, N) and bottoms (O, P, Q) to size.

6 Dry-clamp each drawer to check the fit. Then, glue and clamp each drawer, checking for square by

measuring from corner to corner and adjusting until the opposing diagonal measurements are equal as shown in Photo A, *opposite*.

7 To reinforce the mitered corner joints and add the decorative joinery look, start by building the corner-kerfing jig from the drawing on the *opposite page.*

8 As shown in Photo B on *page 122,* cut a pair of 1/8" spline kerfs 1/4" deep in each front corner of each drawer. For flat bottom kerfs, you'll need to use a blade with a flat-top or triple-chip grind.

9 Cut two pieces of 3/8"-wide 1/8"-thick solid cherry stock to 24" long each. Now, crosscut 48 1/8 × 3/8 × 3/4" cherry splines from the strips. Cut several corner blocks to the size shown on the Drawer drawing. Glue the splines in the kerfs using a corner block on the inside of the drawer as shown in

continued

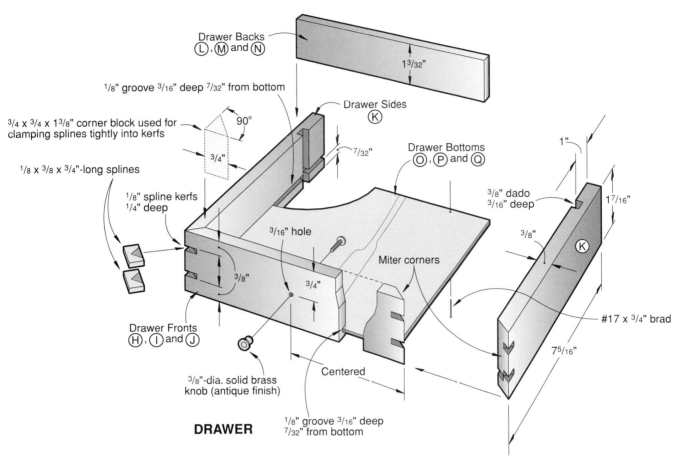

Drawer Backs
(L), (M) and (N)

1³/₃₂"

¹/₈" groove ³/₁₆" deep ⁷/₃₂" from bottom

90°

Drawer Sides
(K)

³/₄ x ³/₄ x 1³/₈" corner block used for clamping splines tightly into kerfs

³/₄"

⁷/₃₂"

Drawer Bottoms
(O), (P) and (Q)

1"

¹/₈ x ³/₈ x ³/₄"-long splines

³/₈" dado
³/₁₆" deep

1⁷/₁₆"

¹/₈" spline kerfs
¹/₄" deep

³/₁₆" hole

³/₈"

(K)

³/₈"

Miter corners

³/₄"

Drawer Fronts
(H), (I) and (J)

#17 x ³/₄" brad

³/₈"-dia. solid brass knob (antique finish)

Centered

7⁵/₁₆"

¹/₈" groove ³/₁₆" deep
⁷/₃₂" from bottom

DRAWER

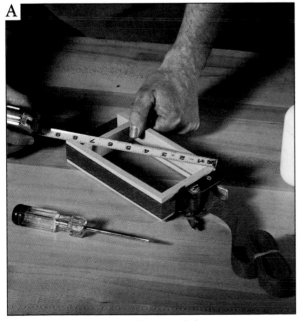

A

Measure diagonally and adjust until the measurements are perfectly equal, to ensure square drawers.

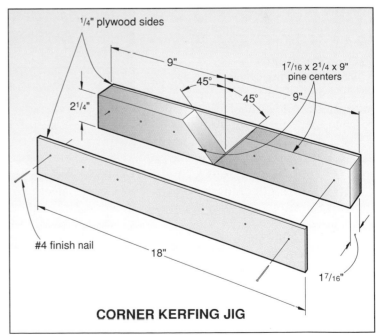

¹/₄" plywood sides

9"

1⁷/₁₆ x 2¹/₄ x 9"
pine centers

45°

45°

9"

2¹/₄"

#4 finish nail

18"

1⁷/₁₆"

CORNER KERFING JIG

Dresser-Top Delight

continued

photo C. (For an even distribution of pressure, we placed a small piece of scrap stock between the clamp head and the splines.) Check that the splines bottom out in the kerfs. If they don't, you'll have an unsightly glue joint later.

10 Trim the splines to within about $\frac{1}{16}$" of the drawer front and sides, and then sand them flush.

ADD THE GUIDES FOR EVENLY SPACED DRAWERS

1 Cut the drawer guides (R, S, T) to size. The width of the guides needs to be equal to the distance between the drawer sides (K) minus $\frac{1}{32}$".

2 Slide the top three drawers into place and position the drawers so you have an equal gap between the drawers and the sides (A) and the same gap between the drawers. Since the drawers were made to fit tight, you'll have to sand the outside faces of the drawer sides to create the $\frac{1}{32}$" gaps. (For consistent-sized gaps, we placed cereal-box cardboard between the drawers and between the drawers and jewelry box side panels. We used sandpaper wrapped around a block of wood to sand the outside surfaces of the drawers evenly.)

3 Working from the back of the case, slide the drawer guides in place. The guides should be $\frac{3}{8}$" (the same as the thickness of the drawer fronts) from the front of the case. Check for equal gaps and that the drawer fronts are flush with each other and the front of the case. Clamp (no glue) the guides in place from the back of the case. Push the drawers out the front without moving the guides. Use a sharp pencil to mark the location of the guides on the shelves. Using a few drops of glue (no need to overdo it) glue the guides in place. *Immediately reposition the drawers and spacers to*

Using the corner kerfing jig for support, cut a pair of kerfs in the front corners of each drawer.

Use corner blocks to prevent denting the inside of the drawers when pulling the splines snug into the kerfs.

verify that the guides are correctly positioned. Evenly spaced drawers depend on properly positioned guides, so take your time. Repeat the process to install the remaining guides.

4 After gluing all the guides in place, install the drawers, and mark the knob-hole centerpoints where shown on the Front View drawing. Note that the holes in drawer fronts I and J align. The machine screws supplied with the

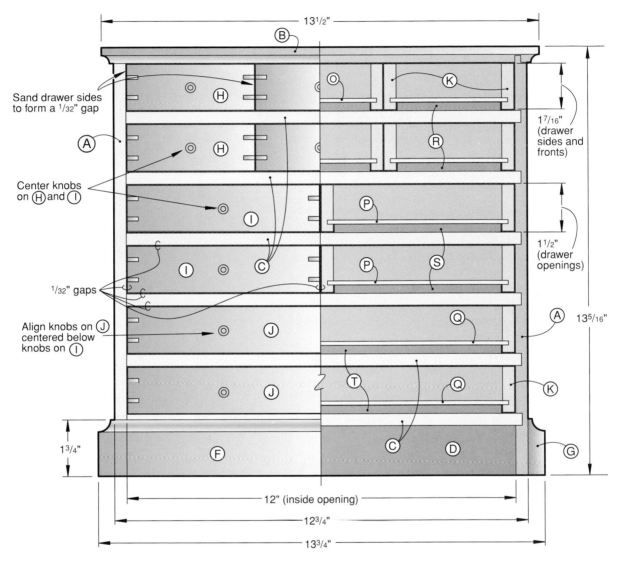

Sand drawer sides to form a ¹/₃₂" gap

Center knobs on Ⓗ and Ⓘ

¹/₃₂" gaps

Align knobs on Ⓙ centered below knobs on Ⓘ

13¹/₂"

1⁷/₁₆" (drawer sides and fronts)

1¹/₂" (drawer openings)

13⁵/₁₆"

1³/₄"

12" (inside opening)

12³/₄"

13³/₄"

FRONT VIEW

knobs are a bit too long for the ³/₈"-thick drawer fronts, so trim each screw accordingly.

5 Remove the drawers from the assembled jewelry box. Finish-sand as necessary and apply the finish. (We applied a coat of Minwax Natural Danish Oil and let it sit for 30 minutes. While still wet, steel-wool the finish, and wipe dry with a clean cloth. Let this dry completely, and repeat the process with a second

coat of finish to completely seal the wood. For added luster, apply a couple coats of 100% tung oil following the directions on the can.)

BUYING GUIDE

• **Hardwood kit.** All the individual pieces shown in the Cutting Diagram cut slightly oversized in length and width from the thicknesses listed in the Bill of Materials. Available in cherry, stock no. W75C, or walnut, stock no.

W75W. For current prices contact Heritage Building Specialties, 205 North Cascade, Fergus Falls, MN 56537. Or call 800/524-4184 to order.

• **Knobs.** For current prices of fourteen solid-brass antique-finish knobs, stock no. 34546, contact The Woodworkers' Store, 21801 Industrial Blvd., Rogers, MN 55374-9514. Or call 800/279-4441 to order.

Saint Nick

A decorative holiday puzzle

Here's a rendition of jolly old Saint Nick you can scrollsaw in a jiffy. Simple carving, woodburning, and painting make this stand-up Santa puzzle as festive as yuletide itself.

Trace the *black* cutting lines from the full-sized pattern *opposite page* onto a ¾×6×10" piece of basswood or clear pine. Place the pattern bottom along one end of the stock.

Scrollsaw around the outside pattern line, starting the cut from a bottom corner. A #5 blade (.038×.016" with 12.5 teeth per inch) will handle the cutting. Next, cut the pieces apart. Here's one way to proceed: Starting at the bottom of the basket, saw up the pattern line on Santa's left arm, returning to the outside at his left shoulder. From there, cut around the line separating his hair from his face, exiting at his right shoulder. Then, cut out Santa's right arm, followed by the tree. Finally, separate the lower portion along the beard line.

Sand the parts as necessary, removing any fuzz along the cuts. Trace the *blue* and *orange* detail lines from the pattern onto the cut-out pieces.

Carve a shallow groove along each *orange* line. Carve the coat hem and collar line on Santa's back. Also on the back, carve the lines on the tree, the left hand and basket, and the fur trim on the hat and sleeves. An incision about ¹⁄₁₆" wide made with a woodcarver's no. 12 V-tool would be ideal. The grooves don't need to meet any particular specification—they just help define features for painting.

Woodburn the *blue* lines, using a skew tip on your woodburning pen. Texture the tree by woodburning fine, sweeping lines on each layer, following the style shown on the top

layer on the pattern. Woodburn both sides of the tree. For Santa's basket, woodburn the basket-weave pattern on both sides and the exposed edge. Don't texture the basket handle.

Paint the completed Santa cutout with acrylic artist's colors. Paint the exposed edges, but not the joining edges. Kansas City artist Fern Weber, who designed our Santa, used these

Delta Ceramcoat colors, shown on the pattern: Berry red (BR), Indiana rose (IR), Copen blue (CB), Green isle (GI), Wedgwood green (WG), Dusty purple (DP), Burnt sienna (BS), Burnt umber (BU), Light ivory (LI), White (WT).

To create the fur texture, add a bit of gel thickener to the white paint. (The paints and thickener are

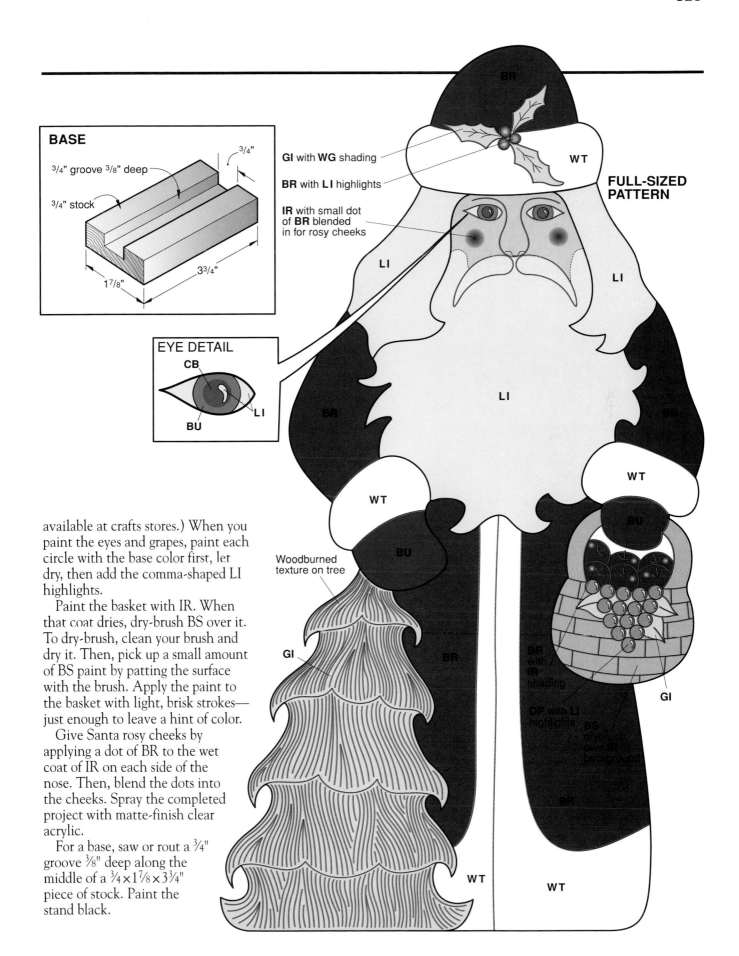

BASE

3/4" groove 3/8" deep

3/4" stock

3/4"

1 7/8"

3 3/4"

GI with **WG** shading

BR with **LI** highlights

IR with small dot of **BR** blended in for rosy cheeks

FULL-SIZED PATTERN

WT

LI

LI

LI

BR

BR

WT

BU

WT

BU

Woodburned texture on tree

GI

BR

GI

BR with IR shading

DP with LI highlights

BS drybrush over IR background

GI

BR

WT

WT

EYE DETAIL

CB

LI

BU

available at crafts stores.) When you paint the eyes and grapes, paint each circle with the base color first, let dry, then add the comma-shaped LI highlights.

Paint the basket with IR. When that coat dries, dry-brush BS over it. To dry-brush, clean your brush and dry it. Then, pick up a small amount of BS paint by patting the surface with the brush. Apply the paint to the basket with light, brisk strokes—just enough to leave a hint of color.

Give Santa rosy cheeks by applying a dot of BR to the wet coat of IR on each side of the nose. Then, blend the dots into the cheeks. Spray the completed project with matte-finish clear acrylic.

For a base, saw or rout a ¾" groove ⅜" deep along the middle of a ¾ × 1⅞ × 3¾" piece of stock. Paint the stand black.

Tour-De-Force Tureen

You'll draw raves with this redwood-burl bowl

Looks great! Saves wood! For many turners, advantages like these offset the extra effort that goes into building a staved blank. See for yourself, and feel the pride of accomplishment as you create this lovely lidded bowl, another great design from the lathe of WOOD® magazine's art director and master turner Lee Gatzke.

Bowl turning can be so straightforward. Just attach a solid block of wood to the lathe, put a gouge to it, and presto! A bowl! Why, then, would anybody go to the effort of cutting a lot of wedge-shaped pieces of wood (*staves*) and gluing them together to construct a bowl blank?

Here's one practical reason: A lot less wood goes to waste. A staved blank starts out nearly round, and it's already hollow, so you generate fewer chips while roughing out your vessel. That, along with the fact that you can use small pieces of stock—even scraps—makes staved construction economical. And, of course, you aren't limited to making a bowl only as large as the biggest chunk of wood you can find.

You'll find aesthetic advantages, too. Building up a blank from staves lets you display face grain or a special figure all around a turning. And, it gives you almost endless opportunities to mix different woods for color, texture, and grain patterns that will make your finished piece unique.

This gorgeous lidded bowl makes a strong case for the advantages of staved construction. The beautiful figure of the redwood burl shows from every angle, set off by stripes of maple veneer. As for economy, this bowl's dimensions would require a solid block of heroic proportion (and cost).

BUILD THE BOWL BLANK

To construct the bowl blank, first cut 16 staves like the one shown on *page 128, top left,* from 2"-thick stock. The wide faces of the staves become the outside of the blank. Cut 16 pieces 2×2×6", then bevel-rip the edges. Saw a few pieces of scrapwood the same size for test cuts.

You'll need a carrier like the one on *page 128, top right,* to bevel-rip the edges safely and accurately. Build one from scrapwood, using the dimensions in the drawing. (We first built a three-sided tray with inside dimensions of 2×2¼×6", using simple nailed and glued butt joints. Then, with the tablesaw, we trimmed the carrier to 1⅞" wide at an 11¼° angle where shown. When you cut the jig to width, leave the saw fence in position to rip the staves.)

Saw the bevel on both edges of each stave. Place each blank into the carrier with the best face down. (Test-cut four scrapwood staves to check the jig's accuracy. When you put the four together, they should form a 90° angle between their outside edges.)

After sawing the staves, cut 16 pieces of maple veneer 2" wide by 6" long. Be sure to use solid veneer for these pieces, not self-adhesive or paper-backed veneer. Only the thin edge will show in the completed turning, so you don't need veneer with a fancy figure, either.

Stand the staves on your bench to form a cylinder, then slip a piece of veneer into each joint. Cinch up the assembly with a band clamp, and inspect all joints for tightness. Check the outside diameter now, too. It should measure somewhere between $9\frac{5}{8}$" and $9\frac{3}{4}$", depending on the thickness of your veneer strips.

Glue together the 32 components of the blank with slow-set epoxy. Alternate veneer strips and staves. As you build the glue-up, stand it on a piece of waxed paper on your benchtop. When fully assembled, wrap the built-up blank with waxed paper, and secure with band clamps.

Lay waxed paper and a $\frac{3}{4}\times10\times10$" piece of plywood on top, and clamp the assembly to your bench to keep the ends even. With a square, check the glue-up to be sure it is standing up straight. After the glue cures, scrape the squeeze-out from the sides and ends.

To square the ends of the glued-up cylinder, construct the jig shown on *page 128, bottom right,* for your tablesaw. Set your saw's miter gauge at 90° to the blade, then position the jig so you can saw $\frac{1}{8}$" or slightly less from the end of the blank. Screw the jig to the miter gauge.

Place the cylinder in the jig, and saw the end, rotating the cylinder between cuts. When you have cut away around one end, turn the cylinder end-for-end in the jig, slide a $\frac{1}{8}$"-thick packing of hardboard between the back of the jig and the square-cut end of the cylinder, and trim the other end.

IT'S TIME TO TURN THE BODY

Mount a 10"-diameter wooden auxiliary faceplate about 1" thick on your lathe's metal faceplate. True the face and edge of the auxiliary faceplate. Center the end of the blank that will be the top of the bowl on the faceplate, then glue it into place. Here's one way to center the blank: Turn a recess in the faceplate $\frac{1}{16}$" deep and the same diameter as the blank. Glue the blank into the recess.

continued

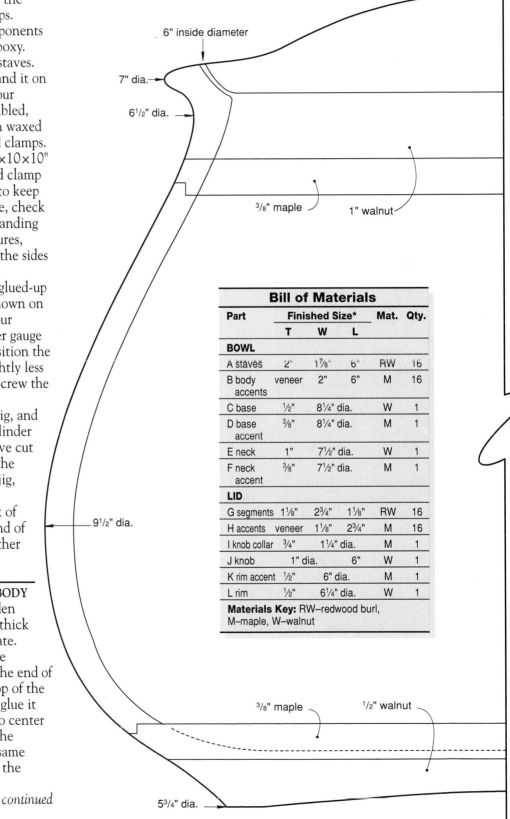

FULL-SIZED HALF PATTERN

6" inside diameter

7" dia.

$6\frac{1}{2}$" dia.

$\frac{3}{8}$" maple

1" walnut

$9\frac{1}{2}$" dia.

$\frac{3}{8}$" maple

$\frac{1}{2}$" walnut

$5\frac{3}{4}$" dia.

Bill of Materials				
Part	**Finished Size***			**Mat.**	**Qty.**
	T	**W**	**L**		
BOWL					
A staves	2"	$1\frac{7}{8}$"	6"	RW	16
B body accents	veneer	2"	6"	M	16
C base	$\frac{1}{2}$"	$8\frac{1}{4}$" dia.		W	1
D base accent	$\frac{3}{8}$"	$8\frac{1}{4}$" dia.		M	1
E neck	1"	$7\frac{1}{2}$" dia.		W	1
F neck accent	$\frac{3}{8}$"	$7\frac{1}{2}$" dia.		M	1
LID					
G segments	$1\frac{1}{8}$"	$2\frac{3}{4}$"	$1\frac{1}{8}$"	RW	16
H accents	veneer	$1\frac{1}{8}$"	$2\frac{3}{4}$"	M	16
I knob collar	$\frac{3}{4}$"	$1\frac{1}{4}$" dia.		M	1
J knob	1" dia.		6"	W	1
K rim accent	$\frac{1}{2}$"	6" dia.		M	1
L rim	$\frac{1}{2}$"	$6\frac{1}{4}$" dia.		W	1

Materials Key: RW—redwood burl, M—maple, W—walnut

Tour-De-Force Tureen

continued

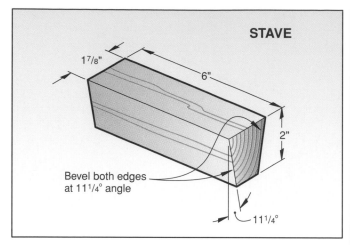

STAVE

1⅞"
6"
2"
Bevel both edges
at 11¼° angle
11¼°

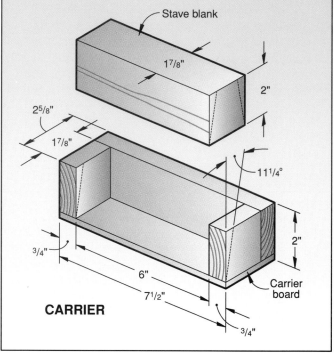

Stave blank
1⅞"
2"
2⅝"
1⅞"
11¼°
¾"
6"
7½"
CARRIER
¾"
Carrier
board

Turn the cylinder to the bowl's largest finished diameter, 9½". True the end with a light scraper cut. Then, following the profile shown on *page 127*, turn the outside of the bowl to shape, using a bowl gouge.

Turn the inside to correspond to the outside shape. As you work, gauge the wall thickness with a caliper at several points to ensure uniformity. Don't attempt to make your bowl too thin—a wall thickness of ¼" is ideal.

Form a ⅛ × ⅛" rabbet on the end of the bowl body to receive the base, as shown on the template. Make the cut by coming straight in from the end of the turning with your parting tool. Dismount the faceplate and workpiece from your lathe. Remove the metal faceplate from the back of the auxiliary faceplate.

ADD THE TWO-TONE BASE

Attach a new auxiliary faceplate to your metal faceplate, and true the face and edge. Glue together an 8¼"-diameter circle of ½"-thick walnut and one of ⅜"-thick maple. After the glue dries, center and glue the walnut side of the lamination onto the wooden faceplate.

Measure the outside diameter of the body at the bottom. Add ¼" to that measurement, then turn the base lamination to that size.

Next, measure the inside diameter of the rabbet in the base of the body. Turn a rabbet on the maple face of the base lamination to mate with the one in the body. To ensure a tight joint, check the fit frequently as you work.

With the base lamination still mounted on the lathe, apply glue to the rabbet and assemble the body and base. Make sure the body is

square to the base, then slide the lathe tailstock up to clamp the parts together.

After the glue has dried, turn the outside of the base to profile. Part the auxiliary faceplate from the top of the bowl. Inside the bowl, cut a

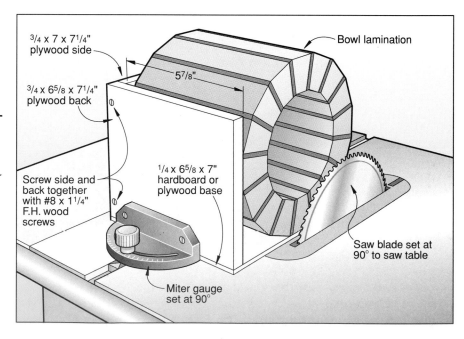

¾ x 7 x 7¼"
plywood side
¾ x 6⅝ x 7¼"
plywood back
Screw side and
back together
with #8 x 1¼"
F.H. wood
screws
¼ x 6⅝ x 7"
hardboard or
plywood base
Miter gauge
set at 90°
Bowl lamination
5⅞"
Saw blade set at
90° to saw table

slight curve as indicated by the dotted line on the template. Be careful not to cut into the rabbet joint, which would weaken it.

THE NECK COMPLETES IT

Trim the turning to 6½", measuring from the bottom of the base to the top edge of the body. Now, form a ⅛ × ⅛" rabbet in the top edge, as you did for the base.

Dismount the faceplate and workpiece from the lathe. Remove the metal faceplate from the back of the auxiliary faceplate, first marking both so you can rejoin them later with the same orientation. Attach a new auxiliary faceplate to the metal faceplate, and true it.

Now, laminate stock for the neck. Glue a 7½"-diameter circle of 1"-thick walnut to a same-sized circle of ⅜"-thick maple. Center and glue the walnut side onto the auxiliary faceplate

Turn the neck lamination to 7" diameter. Then, following the template, rough-shape the neck inside and out. Allow an extra ⅛" or so on both surfaces. As you did with the base, turn a rabbet in the maple side of the neck lamination to mate snugly with the body.

To glue the neck piece to the body, remove the metal faceplate from the back of the neck turning, and reattach it to the bottom of the bowl. Mount the bowl on the lathe, apply glue to the rabbet, and position the neck turning. Again, clamp with the tailstock.

After the glue dries, part off the auxiliary faceplate at the top of the neck. Complete shaping the neck, blending it into the bowl body inside and out. Take care not to cut into the rabbet joint. Also, shape the neck opening to receive the lid.

Sand the inside with a flap sander while the lathe is running. Also with the lathe running, sand with progressively finer grits of hand-held

sandpaper, ending with 320 grit. Similarly sand the outside, beginning with a disc sander instead of the flap sander.

Apply satin Deft lacquer or other clear finish. (We applied three coats with a cloth and sprayed two final coats, sanding between them with 320-grit sandpaper.) Part the bowl from the lathe. Sand and finish the bottom.

NOW, PUT A LID ON YOUR BOWL

Cut 16 segments for the lid from 1⅛"-thick redwood burl. To saw the wedge-shaped pieces safely and accurately, build a sliding-table jig for your tablesaw, following the illustration *below*.

For the segments, start with 1⅛ × 3 × 18" stock (or as long as possible). Joint one edge, then rip the piece to 2¾" wide—both edges must be straight and parallel for accuracy. Prepare a same-sized piece of scrapwood.

Using the sliding table, saw one end of your scrapwood at 11¼°. Now, to cut the first wedge, flip the piece over, placing the opposite edge against the jig's fence, as shown by Cut 2 in the illustration on *page 130, top left*. Flip the stock again, align the short corner with the right side of the kerf as shown, and cut the next wedge. Flip the wood, and repeat.

After you have cut four wedges, lay them next to each other on a smooth surface. The outside edges should form a 90° angle. If not, recheck the angle of the jig's fence, adjust as needed, and try again. When you have adjusted the jig to cut accurately, saw the 16 redwood wedges, using the same alternating-edge scheme.

ASSEMBLE THE LID BLANK

Cut 16 pieces of maple veneer 1⅛ × 2¾". Arrange the 32 parts into a circle, alternating maple slices and redwood wedges. Tighten a band

clamp around the assembly, and check the joints for tightness.

Then, glue with slow-set epoxy, arranging the segments on a piece of waxed paper on your benchtop. Wrap the outside edge of the glue-up with waxed paper, and clamp. To keep it flat, lay waxed paper and a scrapwood disc on top, and clamp to the benchtop.

Allow the glue to cure, then remove the clamps and scrape off the squeeze-out. Sand both faces, keeping them parallel. With a Forstner bit or holesaw, bore a 1⅛" hole through the center.

Bandsaw a 1¼"-diameter circle from ¾"-thick maple. Bore a ⅝" hole through the center. Now, mount a 6" length of 1" walnut dowel between centers on your lathe. Turn the tailstock end to a ⅝"-diameter tenon ¾" long. Glue the maple ring onto the tenon, keeping the dowel perpendicular to the ring.

Grip the end of the walnut dowel in a 3- or 4-jaw lathe chuck. If you don't have one, attach a 1½"-thick auxiliary faceplate to your lathe's faceplate. Round down the edge. Then, with a gouge, form a centered hole 1" deep to hold the dowel snugly.

Turn the maple ring on the dowel to fit into the 1⅛"-diameter center hole bored through the redwood-and-maple glue-up. Glue the larger part onto the maple ring, pushing it on far enough to align the top of the
continued

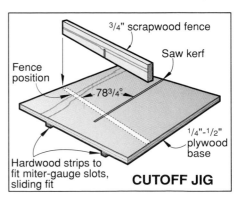

CUTOFF JIG

¾" scrapwood fence
Saw kerf
Fence position
78¾°
¼"-½" plywood base
Hardwood strips to fit miter-gauge slots, sliding fit

Tour-De-Force Tureen

continued

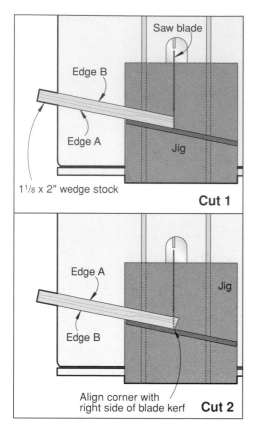

1¹/₈ x 2" wedge stock

Edge B
Edge A
Saw blade
Jig

Cut 1

Edge A
Edge B
Jig

Align corner with
right side of blade kerf **Cut 2**

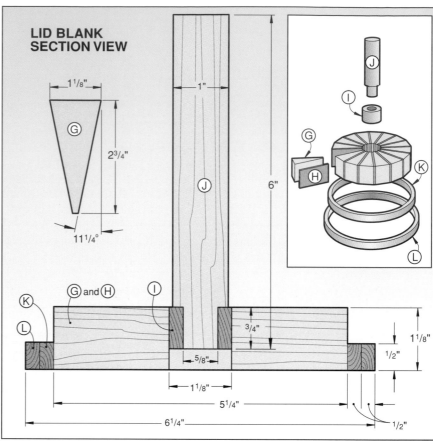

**LID BLANK
SECTION VIEW**

maple ring flush with the top of the larger piece. (See the Lid Blank illustrations *above.*) Keep the larger circle parallel to the faceplate as you glue it into position. After the glue cures, turn the large disc to 5¼" outside diameter.

On another faceplate, attach a 6"-diameter, ¾"-thick auxiliary faceplate. Bandsaw a 6" circle from ½" maple, and glue it to the auxiliary faceplate. Turn the outside to 5¾" diameter, keeping the edge perpendicular to the face. Cut in from the face with the parting tool to form a 5¼" inside diameter. Fit this ring (K) around the assembly where shown.

In the same manner, start with an auxiliary faceplate and a ½"-thick walnut circle, both 6½" in diameter.

Turn the outside diameter to 6¼", and the inside diameter to 5¾". Fit this ring (L) around the maple ring. This completes the blank for the lid.

Turn the lid to the template profile. Turn the outside first. Match the outside contour on the inside, keeping a wall thickness to match your bowl. Turn the outer rim to fit the bowl opening. Sand the lid inside and out.

Form the knob on top last—you need the strength of the full dowel to support the lid during turning. Part off the lid at the top of the knob. Finish the lid to match the bowl.

TOOLS AND SUPPLIES

• **Stock**
  Approximately 3 bd. ft. redwood burl and 2 sq. ft. of maple veneer,

⅜" maple, ½" walnut, 1" walnut (See the Bill of Materials for sizes, and the Buying Guide for our redwood-burl source.)
• **Lathe equipment**
  Two 4–6" metal faceplates, drive center, revolving tail center
• **Tools**
  ⅛" parting tool, ¼" and ½" bowl gouges, ¼" spindle gouge
• **Speeds**
  Roughing: 800–1000 rpm
  Turning and sanding: 1200–1600 rpm

BUYING GUIDE

• **Redwood burl, maple veneer.** For current prices of redwood burl and maple veneer for one bowl, contact One Good Turn, 3 Regal St., Murray, UT 84107, 801/266-1578.

Decorating Accessories

Often times smaller, more decorative projects can make the difference in the warmth and feel of a home. With this in mind, we've selected a few decorating projects from our 1994 magazines that will add beauty to areas both inside and outside your home.

Show-Stopping Picture Frames

Custom-cut your own

E very year we receive numerous letters from readers interested in framing their own artwork or photos. The reason? Prices for quality picture frames have soared through the roof. But by building one or both of these good-looking hardwood frames, you can pocket a lot of change–and learn some practical woodworking techniques to boot. Come on–let's get started.

TRUE-TO-TRADITION PICTURE FRAME

Note: *The larger, more traditional frame shown here was miter-cut to fit a 14×20" piece of artwork. For other sizes, you'll need to change the frame size.*

We made the inside maple frame first so we could size it around our artwork. Then, we cut and fit the outer walnut frame around the maple frame.

HERE'S HOW TO MAKE THE MAPLE INSIDE FRAME

1 From ⅝" maple (we planed thicker stock to this thickness), cut two pieces each measuring 1½" wide by 42" for the inside frame parts (A). Note that the dimensions given in the Bill of Materials are for the *initial size* of the lineal stock. (To machine the stock for the four inside maple frame pieces, we found it easier to work with two 42"-long pieces rather than one 84"-long strip.) Cut or rout a ¼" rabbet ½" deep along one edge of each piece. See the Section View Detail accompanying the Exploded View drawing for reference.

2 Mark the miter-cut locations on the maple (we used a combination square), and check the marked cutlines against the artwork to verify the length of the pieces. Adjust if necessary. Miter-cut the four maple pieces (A) to length. (You can get one 22½" piece and one 16½" piece from each 42"-long strip. We test-cut scrap first to verify an accurate 45° angle.)

3 Glue and clamp the pieces together, checking the frame for square and ensuring that the top and rabbeted surfaces are flush.

4 Later, remove the clamps and remove any dried glue from the corners. Then, finish-sand both surfaces of the maple frame.

NOW, FOR THE WALNUT OUTER FRAME

1 For the walnut frame, cut two pieces to $3/4 \times 1 1/2 \times 52$" for parts B, two pieces to $1/2 \times 1 \times 52$" for parts C, and two pieces to $1/2 \times 1/2 \times 52$" for frame parts D.

2 To contour the front surface of the B strips, bevel-rip one face of each at 16° where shown on Step 1 of the three-step drawing on the top of *pages 134 and 135*.

3 Position the blade on your tablesaw $1/16$" above the surface of the saw table. Following Step 2 of the drawing, clamp a straightedge to your saw table. Slowly feeding the stock against the straightedge and into the blade, cut a shallow cove along each B strip as shown in Photo A and to the shape shown in the Section View Detail.

4 Using Step 3 of the drawing as a guide, use a palm sander to sand the areas noted to finish shaping the profile of the B strips.

5 With the back edges flat, glue and clamp one C strip and one D strip to one of the contoured center strips (B) where shown in the Section View Detail. Immediately remove any excess glue from the front contoured edge of the frame blank. Repeat the process to laminate the remaining three pieces for the outer frame.

6 Cut or rout a $1/4$" rabbet $1/4$" deep along the inside edge of the walnut outer frame blanks where shown on the Section View Detail.

7 Miter-cut one short and one long length of frame from each walnut lamination to fit around the maple frame.

8 Glue and clamp the miter-cut walnut pieces around the maple frame. Remove excess glue.

9 Apply a clear finish to the frame, and later add a sawtooth hanger to the back of it. Add the artwork and hang.

continued

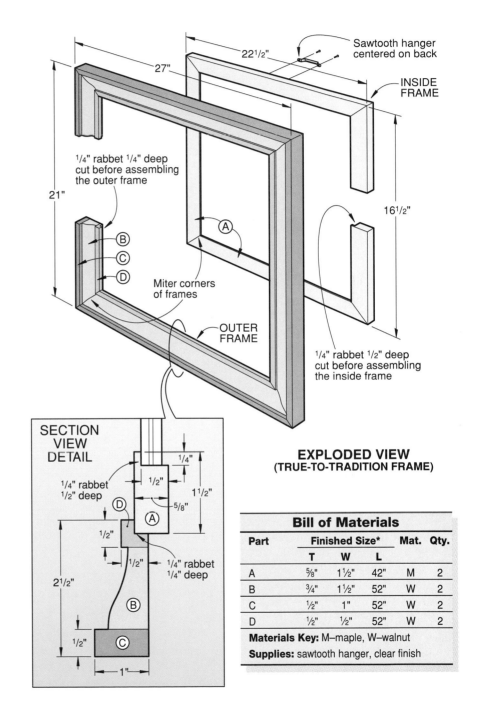

1/4" rabbet 1/4" deep cut before assembling the outer frame

Miter corners of frames

OUTER FRAME

Sawtooth hanger centered on back

INSIDE FRAME

1/4" rabbet 1/2" deep cut before assembling the inside frame

27"   22 1/2"   21"   16 1/2"

EXPLODED VIEW
(TRUE-TO-TRADITION FRAME)

SECTION VIEW DETAIL

1/4" rabbet 1/2" deep

1/4"   1/2"   1 1/2"   5/8"   1/2"   1/2"   1/4" rabbet 1/4" deep   2 1/2"   1/2"   1"

Bill of Materials				
Part	Finished Size*		Mat.	Qty.
	T	W	L		
A	5/8"	1 1/2"	42"	M	2
B	3/4"	1 1/2"	52"	W	2
C	1/2"	1"	52"	W	2
D	1/2"	1/2"	52"	W	2

Materials Key: M–maple, W–walnut

Supplies: sawtooth hanger, clear finish

Show-Stopping Picture Frames

continued

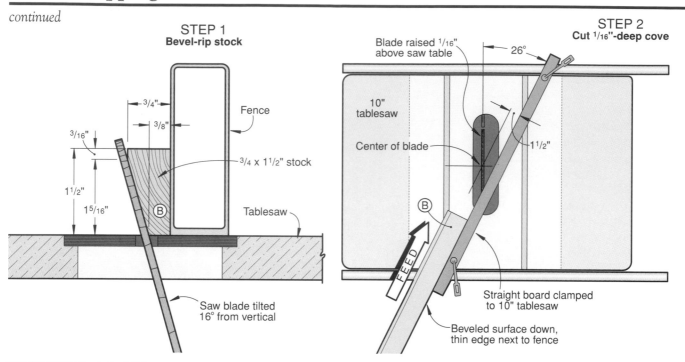

STEP 1
Bevel-rip stock

3/4"
3/8"
3/16"
1 1/2"
1 5/16"
Fence
3/4 x 1 1/2" stock
Tablesaw
B
Saw blade tilted
16° from vertical

STEP 2
Cut 1/16"-deep cove

Blade raised 1/16"
above saw table
26°
10"
tablesaw
Center of blade
1 1/2"
B
FEED
Straight board clamped
to 10" tablesaw
Beveled surface down,
thin edge next to fence

CATCHY-CORNERS PHOTO FRAME

Note: *The frame shown here was made to fit an 8×10" photograph. For other sizes you'll need to change the frame size accordingly.*

FIRST, MACHINE THE STOCK AND ASSEMBLE THE FRAME

1 Cut a piece of 3/4"-thick stock (we used genuine mahogany) to 1 5/8" wide by 50" long.

2 Rout a 1/4" chamfer along one edge of the 50"-long piece.

3 Rout or cut a 1/4" rabbet 3/8" deep along the same edge (opposite face) of the stock. See the Frame Detail for reference.

4 Miter-cut the pieces to length. (We test-cut scrap first to verify an accurate 45° angle.)

5 Glue and clamp the pieces, checking that the frame is square and that the top surfaces are flush.

NOW, CUT THE SPLINE KERFS

1 Attach a miter-gauge extension to your miter gauge. Using the

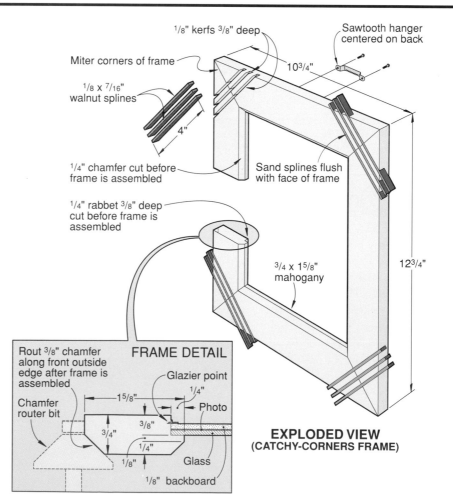

1/8" kerfs 3/8" deep
Sawtooth hanger centered on back
Miter corners of frame
10 3/4"
1/8 x 7/16" walnut splines
4"
1/4" chamfer cut before frame is assembled
Sand splines flush with face of frame
1/4" rabbet 3/8" deep cut before frame is assembled
3/4 x 1 5/8" mahogany
12 3/4"

EXPLODED VIEW
(CATCHY-CORNERS FRAME)

FRAME DETAIL

Rout 3/8" chamfer along front outside edge after frame is assembled
Chamfer router bit
Glazier point
1/4"
Photo
1 5/8"
3/8"
3/4"
1/4"
1/8"
Glass
1/8" backboard

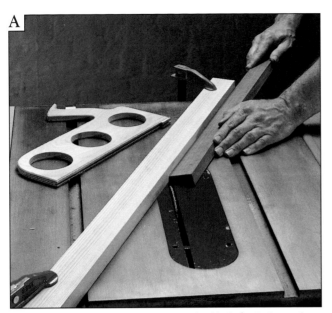

A

Clamp a straightedge in place, raise the blade $\frac{1}{16}$" above the saw surface, and cut a cove in the frame stock.

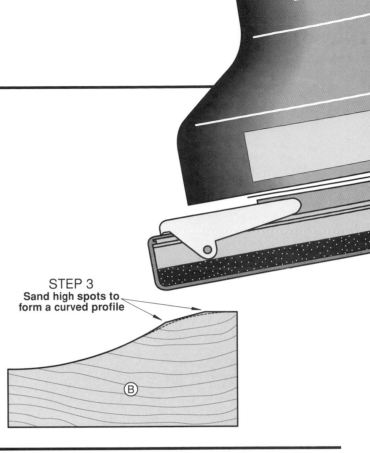

STEP 3
Sand high spots to
form a curved profile

B

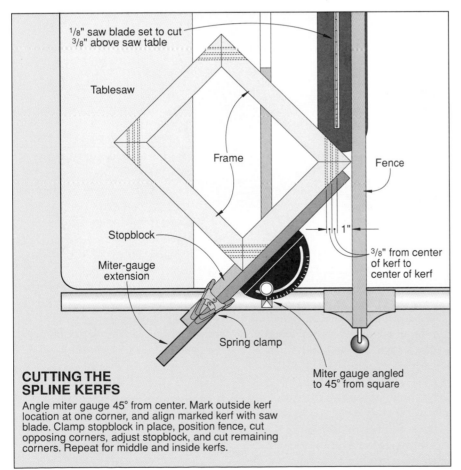

$\frac{1}{8}$" saw blade set to cut
$\frac{3}{8}$" above saw table

Tablesaw

Frame

Fence

Stopblock

Miter-gauge
extension

Spring clamp

1"

$\frac{3}{8}$" from center
of kerf to
center of kerf

Miter gauge angled
to 45° from square

**CUTTING THE
SPLINE KERFS**

Angle miter gauge 45° from center. Mark outside kerf location at one corner, and align marked kerf with saw blade. Clamp stopblock in place, position fence, cut opposing corners, adjust stopblock, and cut remaining corners. Repeat for middle and inside kerfs.

drawing titled Cutting the Spline Kerfs *at left*, mark the kerf locations, position the fence and stop, and cut the kerfs.

2 Plane or resaw maple until its thickness ($\frac{1}{8}$") fits into the kerfs. Next, rip $\frac{7}{16}$"-wide strips, and crosscut 12 pieces 4" long each.

3 Glue one of the $\frac{1}{8} \times \frac{7}{16} \times 4$" splines into each $\frac{3}{8}$"-deep kerf. Check that the splines bottom out in the kerfs. After the glue dries, trim the splines flush with the outside edges of the frame. (We clamped the frame to our workbench, and used a dovetail saw to trim the ends of the splines to within about $\frac{1}{16}$" of the frame edges. Finally, we used a palm-grip sander to sand the splines flush with the frame's edges and front surface.)

4 Rout a $\frac{3}{8}$" chamfer along the outside front edge of the frame.

5 Finish-sand the entire frame and add the finish. From $\frac{1}{8}$" stock (we used plywood) cut the back to fit the rabbeted opening.

Wetlands Visitors

With this lath-art project, you can really make a scene

For quite awhile now, readers have been asking us to do a lath-art design. Until recently, though, we hadn't been able to turn up just the right one. Then, as luck would have it, we discovered the work of Tom Brahill from Toronto, Canada. Here's one of his many crowd-pleasing scenes that you'll want to hang in your home.

START WITH THE THIN STRIPS

1 From ¾"-thick pine stock, rip 27 strips ⅛"-thick by 16½" long. (We used a standard rip blade. The slightly rough surface left by the blade on the strips makes a desirable texture when finishing the project.)

2 Position 22 of the pieces good side down on a flat surface,

holding the pieces tightly together to prevent gaps between the strips.

3 Cut a piece of heavy brown paper such as Kraft or brown wrapping paper to 17" square.

4 Using a 3" paint roller, foam brush, or playing card, spread a thin, even coat of yellow woodworker's glue on the exposed face of the resawn strips. Place the paper against the glued surface, being careful to keep the strips of wood edge-to-edge. (We used a framing square to keep the ends of the pieces flush.) Using a flat piece of wood, level the paper on the wood to remove any air bubbles or unevenness.

Later, after the glue has dried, use a hobby knife to trim the paper even with the edges of the wood.

TRANSFER THE PATTERNS AND CUT THE PIECES TO SHAPE

1 Make four photocopies of the full-sized scene pattern.

2 Using a hobby or an X-acto knife, cut along the solid lines of the first photocopy to cut the paper pattern sections to shape.

3 Using spray adhesive, adhere the sections to the pine (opposite the paper side) in the configuration shown on the Cutting Diagram on the *opposite page*.

4 Using your scrollsaw and a #8 (8 TPI) blade, cut along the *solid* lines of each pattern section as shown in the photo on the *opposite page*.

5 Tape the second full-sized photocopied pattern to your workbench as shown in Step 1 of

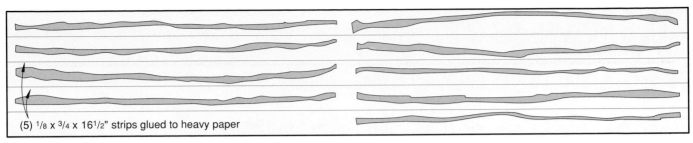

(5) $\frac{1}{8}$ x $\frac{3}{4}$ x 16$\frac{1}{2}$" strips glued to heavy paper

CUTTING DIAGRAM

(22) $\frac{1}{8}$ x $\frac{3}{4}$ x 16$\frac{1}{2}$" strips glued to heavy paper

Assembling the Pieces drawing on *page 41*.

6 Scrollsaw along the *dashed* lines to cut the individual pieces from each section. Number the back side (opposite the photocopied pattern) to match those on your full-sized pattern. Place the individual pieces on the full-sized pattern taped to your workbench.

7 Using the third full-sized pattern, transfer the outlines of the nine willows to the five remaining $\frac{1}{8}$ x $\frac{3}{4}$ x 16$\frac{1}{2}$" pieces of pine. Cut the willows to shape.

PAINT THE PIECES AND ASSEMBLE THE PICTURE

1 Using lacquer thinner to disolve the adhesive, remove the paper patterns from the fronts of the scrollsawed pieces.

2 Using the opening photo and Paint Key as a guide, paint the front side of each pattern piece, and
continued

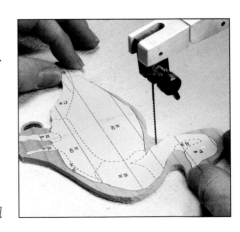

Wetland Visitors

continued

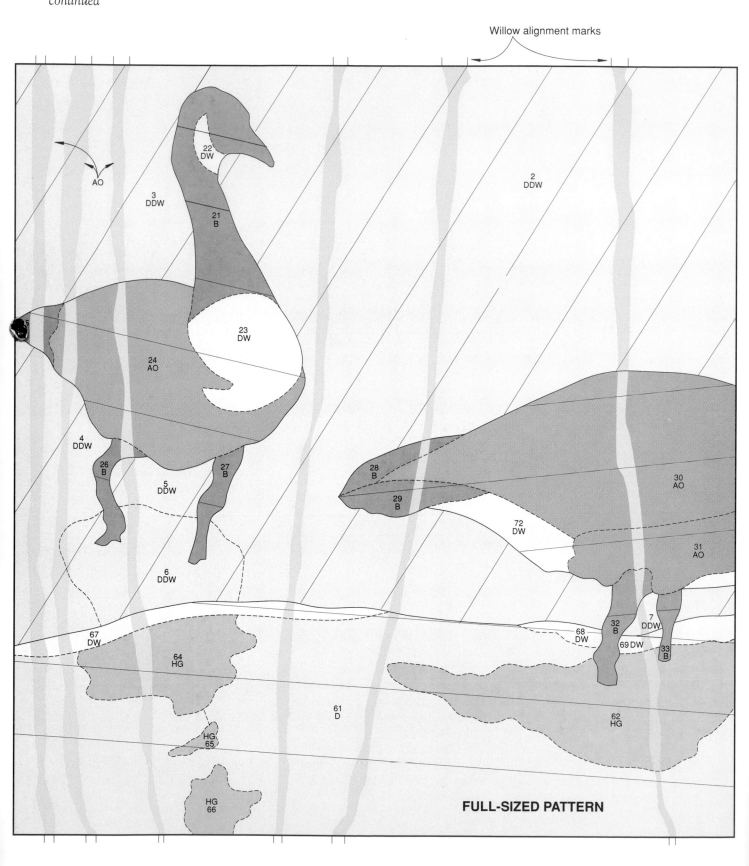

Willow alignment marks

FULL-SIZED PATTERN

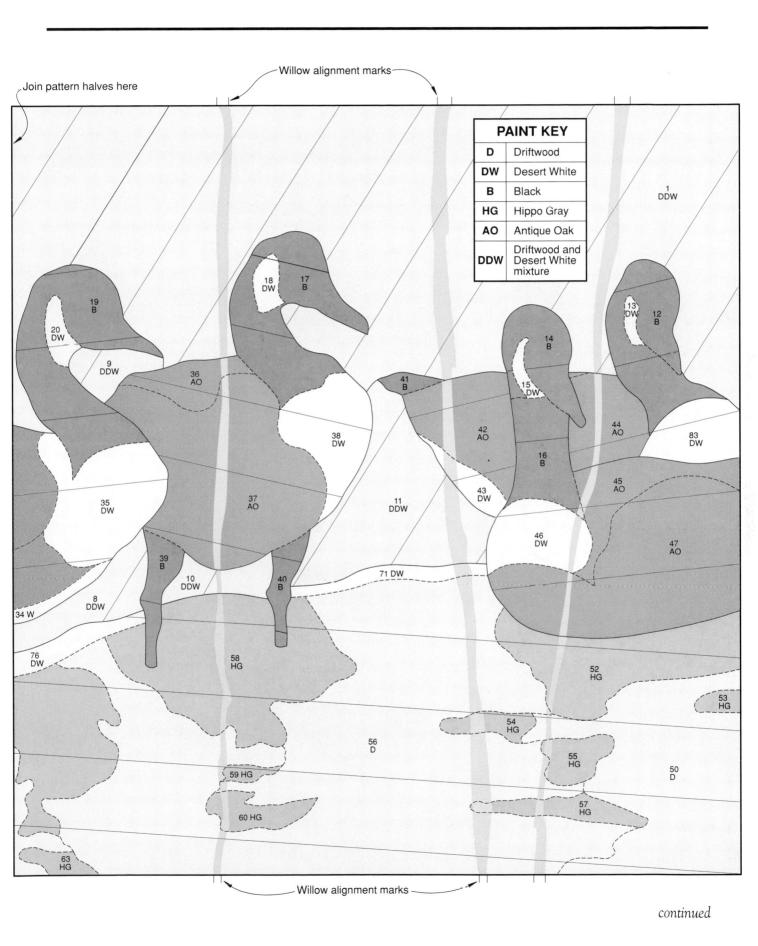

Join pattern halves here

Willow alignment marks

PAINT KEY

D	Driftwood
DW	Desert White
B	Black
HG	Hippo Gray
AO	Antique Oak
DDW	Driftwood and Desert White mixture

Willow alignment marks

continued

Wetland Visitors

continued

immediately wipe off the excess paint for a washed look. You want the grain to show through so don't apply the paint too heavily. Place the painted pieces on top of the reference photocopy taped to your workbench. See the Buying Guide for our source of paints.

3 Paint the front face and edges of the willow pieces the same color as the goose body parts.

4 Cut a piece of ¼" plywood to 10×18". Spray the back side of the fourth photocopy with spray adhesive. Center and adhere the pattern to the ¼" plywood where shown in Step 2 of the Assembling the Pieces drawing.

5 Glue the painted pieces to the pattern adhered to the ¼" plywood as shown in Step 3 of the Assembling the Pieces drawing. (To keep from having to clamp each piece in place, we used instant glue.)

6 Using the alignment marks on the full-sized pattern as guides, glue the willow pieces in place on top of the goose and water pieces.

7 Trim the lath-art scene to 8×15" finished size, making sure the edges are square.

FINISH BY ADDING THE FRAME

1 Cut one piece of pine stock to ¼×⅝×52" for the inner frame parts and one piece to ½×⅞×56" for the outer frame parts.

2 Using the Exploded View drawing for reference, miter-cut the the inner frames pieces to length to fit around the scene. Paint the pieces and glue them to the edges of the scene.

3 Miter-cut the outer frame pieces to length and paint them. Then, glue and nail them in place.

4 Center and attach a sawtooth hanger to the back of the picture for hanging.

BUYING GUIDE

• **Acrylic paint kit.** One 2-oz. bottle each of Delta driftwood, desert white, black, hippo gray, and antique oak. Four full-sized photocopies of the goose scene pattern included. For current price of kit no. WMGEESE, contact Cherry Tree Toys, P.O. Box 369, Belmont, OH 43718. Or call 800/848-4363 to order. A ½"-wide brush and an 8-ounce can of spray adhesive can be purchased for an additional cost.

• **Additional patterns.** For current price of a catalog of lath-art patterns and kits, contact Tom Bramhill Lath Art, 660 Eglington Ave. East, P.O. Box 50101, Toronto, Ontario, M3G 4G1, Canada.

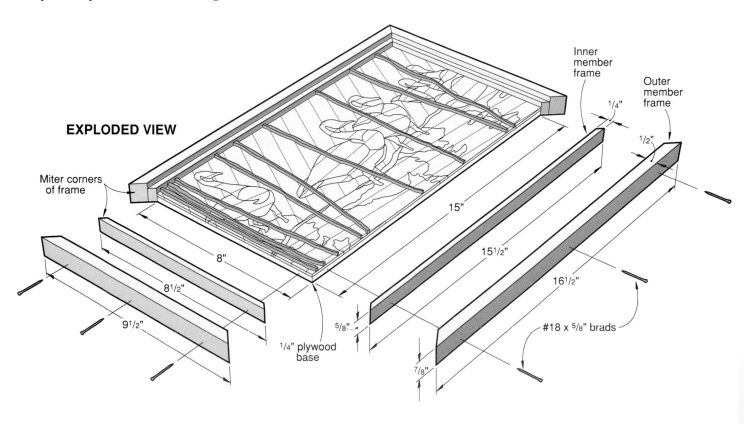

EXPLODED VIEW

Miter corners of frame

Inner member frame

Outer member frame

¼"

½"

15"

15½"

16½"

8"

8½"

9½"

¼" plywood base

⅝"

⅞"

#18 × ⅝" brads

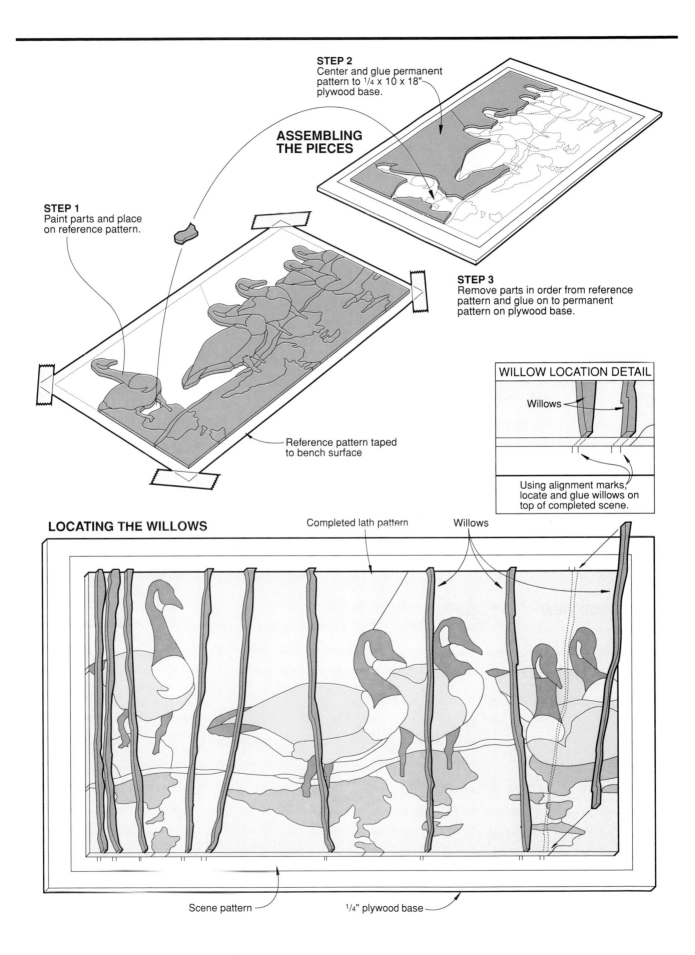

STEP 2
Center and glue permanent pattern to ¹/₄ x 10 x 18" plywood base.

ASSEMBLING THE PIECES

STEP 1
Paint parts and place on reference pattern.

STEP 3
Remove parts in order from reference pattern and glue on to permanent pattern on plywood base.

Reference pattern taped to bench surface

WILLOW LOCATION DETAIL

Willows

Using alignment marks, locate and glue willows on top of completed scene.

LOCATING THE WILLOWS

Completed lath pattern

Willows

Scene pattern

¹/₄" plywood base

Ace-of-Hearts Triplane Whirligig

Modeled after the Fokker triplane flown by the Red Baron in World War I, our whimsical version features a rotating propeller and flapping pilot's scarf. Sturdy construction guarantees that our trusty airplane can fly through years of stormy weather. And its brightly painted surface makes it a pleasure to look at, too.

Note: For joints that will stand up to the extremes of Mother Nature, use Titebond II or Weather-Tite water-resistant glues, slow-set epoxy, or resorcinol glue. Although we used clear pine for our model, Honduras mahogany is an excellent exterior alternative.

START WITH THE LAMINATED FUSELAGE

1 Cut a piece of ¾" stock to 2×20" long for the fuselage (A).

2 Cut a ¼" groove ⅛" deep ¾" from the bottom edge of the piece where shown on the Fuselage drawing. Crosscut two 9⅜"-long fuselage pieces (A) from the 20"-long piece. Next, cut a ¾" dado ½"

deep in each piece where shown on the Fuselage drawing.

3 Cut a piece of pine to ¼×¼×4³⁄₁₆" to plug the crankshaft-tube groove behind the crank opening where shown on the Fuselage drawing.

4 With the edges and ends flush, the filler block in place, and dadoes and grooves aligned, glue and clamp the two ¾" fuselage pieces together.

5 Make a photocopy of the full-sized Side and Top View Patterns, *opposite*. Cut the rectangular paper patterns, and adhere them to the fuselage with spray adhesive, aligning the edges and ends of the paper patterns with those of the laminated fuselage. See the Fuselage drawing for reference.

6 Using a square, transfer the landing-gear strut centerlines to the bottom of the fuselage. Using your drill press, drill a pair of ¼" holes ⁹⁄₁₆" deep into the bottom of the fuselage for the struts. Drill a ⁹⁄₃₂" hole ¾" deep for the pivot tube.

7 Mount a dado blade to your tablesaw, and cut a 2⅛" notch ⅛" deep across the bottom of the fuselage for the bottom wing. Cut a 1¼" notch to the depth shown on the full-sized pattern across the top of the fuselage for the middle wing.

8 Bandsaw the rudder notch in the tail section of the fuselage to size. Drill a ¼" hole in the cockpit. Next, bandsaw the cockpit outline to shape.

9 Following the cutlines on the Top View Pattern, cut the fuselage profile to shape. Save the scraps and use double-faced tape to adhere them back

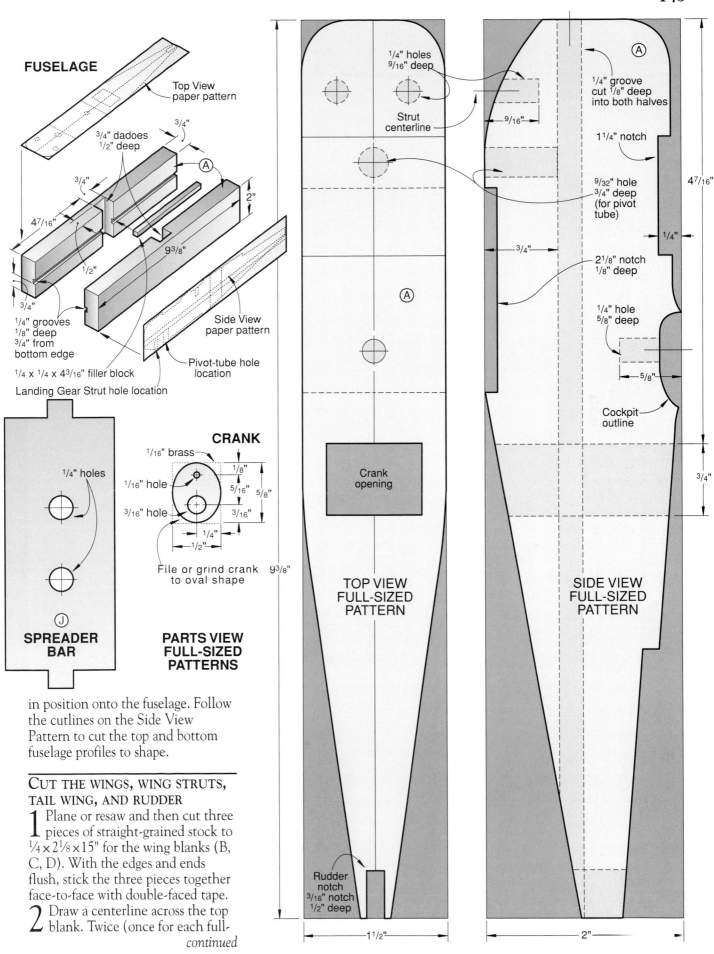

FUSELAGE

Top View paper pattern

3/4" dadoes 1/2" deep

3/4"

Ⓐ

2"

3/4"

4 7/16"

3/4"

1/2"

9 3/8"

1/4" grooves 1/8" deep 3/4" from bottom edge

Side View paper pattern

1/4 x 1/4 x 4 3/16" filler block

Landing Gear Strut hole location

Pivot-tube hole location

SPREADER BAR

1/4" holes

Ⓙ

CRANK

1/16" brass

1/16" hole

3/16" hole

1/8"

5/16"

3/16"

5/8"

1/4"

1/2"

File or grind crank to oval shape

PARTS VIEW FULL-SIZED PATTERNS

9 3/8"

in position onto the fuselage. Follow the cutlines on the Side View Pattern to cut the top and bottom fuselage profiles to shape.

CUT THE WINGS, WING STRUTS, TAIL WING, AND RUDDER

1 Plane or resaw and then cut three pieces of straight-grained stock to 1/4 x 2 1/8 x 15" for the wing blanks (B, C, D). With the edges and ends flush, stick the three pieces together face-to-face with double-faced tape.

2 Draw a centerline across the top blank. Twice (once for each full-
continued

1/4" holes 9/16" deep

Strut centerline

Ⓐ

Crank opening

TOP VIEW FULL-SIZED PATTERN

Rudder notch 3/16" notch 1/2" deep

1 1/2"

9/16"

3/4"

Ⓐ

1/4" groove cut 1/8" deep into both halves

1 1/4" notch

9/32" hole 3/4" deep (for pivot tube)

2 1/8" notch 1/8" deep

1/4" hole 5/8" deep

Cockpit outline

4 7/16"

1/4"

3/4"

5/8"

SIDE VIEW FULL-SIZED PATTERN

2"

Ace-of-Hearts Triplane Whirligig

continued

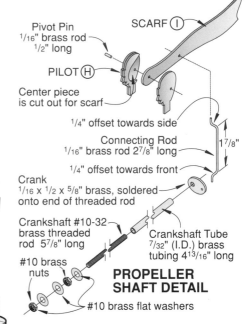

PROPELLER SHAFT DETAIL

Pivot Pin
1/16" brass rod
1/2" long

PILOT (H)

Center piece is cut out for scarf

SCARF (I)

1/4" offset towards side

Connecting Rod
1/16" brass rod 2 7/8" long

1 7/8"

1/4" offset towards front

Crank
1/16" x 1/2 x 5/8" brass, soldered onto end of threaded rod

Crankshaft #10-32 brass threaded rod 5 7/8" long

#10 brass nuts

Crankshaft Tube
7/32" (I.D.) brass tubing 4 13/16" long

#10 brass flat washers

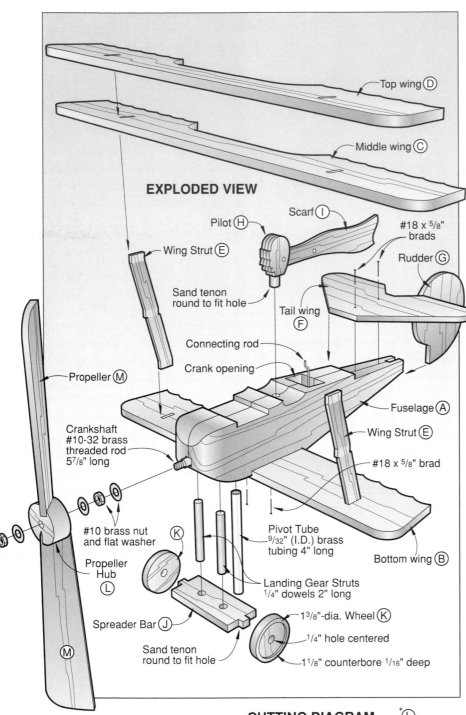

EXPLODED VIEW

Top wing (D)

Middle wing (C)

Pilot (H)

Scarf (I)

Wing Strut (E)

Sand tenon round to fit hole

Connecting rod

Crank opening

Tail wing (F)

#18 x 5/8" brads

Rudder (G)

Propeller (M)

Crankshaft #10-32 brass threaded rod 5 7/8" long

Fuselage (A)

Wing Strut (E)

#18 x 5/8" brad

#10 brass nut and flat washer

Propeller Hub (L)

Spreader Bar (J)

Sand tenon round to fit hole

(K)

Pivot Tube
9/32" (I.D.) brass tubing 4" long

Landing Gear Struts
1/4" dowels 2" long

Bottom wing (B)

1 3/8"-dia. Wheel (K)

1/4" hole centered

1 1/8" counterbore 1/16" deep

(M)

Bill of Materials

Part	Finished Size*			Mat.	Qty.
	T	W	L		
A fuselage	1 1/2"	2"	9 3/8"	LP	1
B btm. wing	1/4"	2 1/8"	12"	P	1
C mdl. wing	1/4"	2 1/8"	14 3/8"	P	1
D top wing	1/4"	2 1/8"	13 3/16"	P	1
E wing struts	1/8"	1 1/2"	4"	P	2
F tail wing	3/16"	3 1/8"	6 1/8"	P	1
G rudder	3/16"	2 3/4"	3 1/4"	P	1
H pilot	3/8"	7/8"	1 5/8"	LP	1
I scarf	3/32"	1"	4 3/4"	P	1
J spreader bar	1/4"	1 1/8"	3"	P	1
K wheels	1/4"		1 3/8"-dia.	P	2
L hub	1/2"		1 1/8"-dia.	P	1
M propeller	1/8"	1 1/4"	5 1/2"	P	2

Materials Key: LP–laminated pine, P–pine.

Supplies: double-faced tape, colored tape (3M plastic tape), solder, flux, 1/4" dowel stock, slow-set epoxy, plus items listed in the Buying Guide.

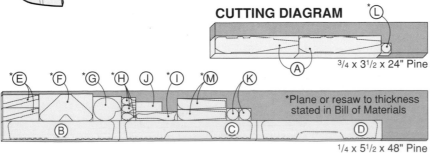

CUTTING DIAGRAM

*(L)

(A)

3/4 x 3 1/2 x 24" Pine

*(E) *(F) *(G) *(H) (J) *(I) *(M) (K)

(B) (C) (D)

*Plane or resaw to thickness stated in Bill of Materials

1/4 x 5 1/2 x 48" Pine

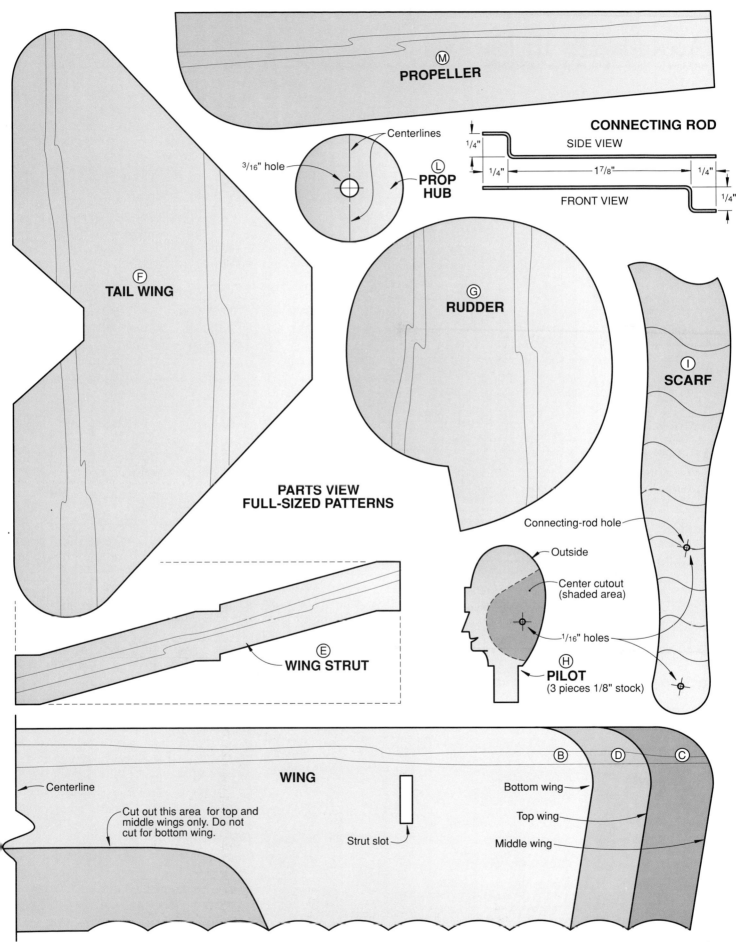

M PROPELLER

Centerlines

$3/16''$ hole

L PROP HUB

CONNECTING ROD

SIDE VIEW

$1/4''$
$1/4''$ — $1^{7}/8''$ — $1/4''$
$1/4''$

FRONT VIEW

F TAIL WING

G RUDDER

I SCARF

**PARTS VIEW
FULL-SIZED PATTERNS**

Connecting-rod hole

Outside

Center cutout
(shaded area)

$1/16''$ holes

H PILOT
(3 pieces 1/8" stock)

E WING STRUT

WING

Centerline

Cut out this area for top and
middle wings only. Do not
cut for bottom wing.

Strut slot

B D C

Bottom wing
Top wing
Middle wing

continued

Ace-of-Hearts Triplane Whirligig

continued

sized trio of half patterns), transfer the wing patterns to the top wing blank, aligning the centerline on the paper pattern to that marked on the top blank. Using a bandsaw or scrollsaw, cut scallops along the back edge of the three taped-together wing blanks.

3 Drill blade start holes, and use a scrollsaw or coping saw to cut the strut slots through all three wing blanks to shape.

4 Separate the wing blanks, transfer the cutlines, and finish cutting the three wings to shape.

5 Transfer the patterns and cut the wing struts (E), tail wing (F), and rudder (G) to shape from the thickness of stock listed in the Bill of Materials.

6 To cut the pilot pieces (H) to shape, stack and tape three 1×4" pieces of 1/8" stock face-to-face. Transfer the pilot outline to the taped-together pieces, and cut the

stack to shape. Separate the three pieces and cut the center section to final shape. See the Propeller Shaft Detail for reference when cutting the pilot center to shape.

7 Cut the 3/32"-thick scarf (I) to shape. Drill the 1/16" connecting rod hole in the scarf where marked. Angle the drill bit up and down slightly when drilling the hole to allow the connecting rod to operate freely when installed later.

LET THE ASSEMBLY BEGIN

1 Glue and nail the bottom wing into its respective notch in the bottom of the fuselage.

2 Insert the struts through the slots in the middle wing. Glue the middle wing/struts assembly to the fuselage. Glue and insert the bottom end of the struts into the slots in the lower wing. Glue the top wing to the top of the struts. Putty any imperfections and sand smooth.

3 Glue the tail wing (F) and rudder (G) in place.

4 Glue the three pilot pieces together with the edges flush. Sand bottom of pilot to fit into 1/4" hole. Drill a 1/16" hole through the lamination.

5 Cut a piece of 1/16"-diameter brass rod to 1/2" long. Insert the rod through the 1/16" holes in the head and scarf. On a hard metal surface, use a ball peen hammer to lightly tap both ends of the brass rod to rivet (mushroom) each end.

ADD THE AXLE, STRUTS, AND WHEELS FOR A SMOOTH LANDING

1 Transfer the outline and the hole centerpoints for the spreader bar (J) to 1/4"-thick stock. Drill the 1/4" holes, and cut the spreader bar to shape.

2 Crosscut two pieces of 1/4" dowel stock to 2" for the landing gear struts. Glue the dowels into the holes in the spreader bar, and into the holes in the bottom of the fuselage.

3 To form the wheels (K), use a compass to mark two 1 3/8"-diameter (11/16" radius) circles on 1/4"-thick stock. Using a 1 1/8" Forstner bit, bore a 1/16"-deep depression at each centerpoint. Then, drill a 1/4" hole through the center of each wheel. Cut the wheels to shape. Sand the tenoned ends of the spreader bar until they fit into the holes in the wheels. Glue the wheels in place.

MACHINE THE PROPELLER FOR MAXIMUM POWER

1 To form the propeller hub (L), start by cutting a 1 1/8"-diameter disc from a piece of 1/2" pine. Next, drill a 3/16" hole through the hub at the centerpoint.

2 Attach a wood extension to your miter gauge, and follow the 5 steps on Notching the Propeller Hub drawing *opposite* and shown in the photo *opposite* to cut the kerfs.

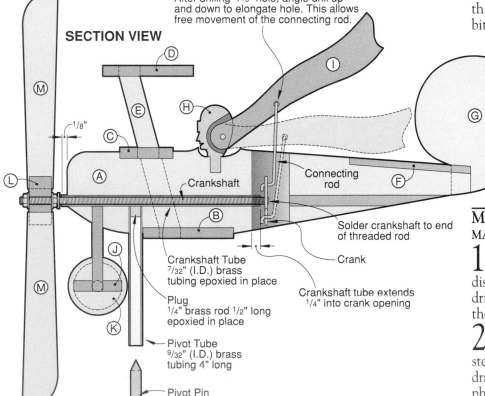

SECTION VIEW

After drilling 1/16" hole, angle drill up and down to elongate hole. This allows free movement of the connecting rod.

Crankshaft

Connecting rod

Solder crankshaft to end of threaded rod

Crank

Crankshaft tube extends 1/4" into crank opening

Crankshaft Tube 7/32" (I.D.) brass tubing epoxied in place

Plug 1/4" brass rod 1/2" long epoxied in place

Pivot Tube 9/32" (I.D.) brass tubing 4" long

Pivot Pin 1/4" brass rod sharpened on top end

3 Cut the propeller blades (M) to shape. Next, glue or epoxy them to the propeller hub (L).

THE METAL PARTS CONVERT WIND ENERGY TO MOTION

1 Using the Propeller Shaft detail accompanying the Exploded View drawing and the Section View drawing for reference, cut all the brass parts except the crank to length. Drill the holes in one end of the crank blank (see the Parts View Patterns for reference). It's safer drilling the holes in a long piece rather than the ⅝"-long finished piece. Crosscut the crank to length and file to shape.

2 Slide and epoxy the ⁷⁄₃₂" (I.D.) brass tubing 4¹³⁄₁₆" long (crankshaft tube) into the fuselage until it protrudes into the crank opening where shown on the Section View drawing. Slide the #10-32 brass threaded rod (crankshaft) into the brass crankshaft tube.

3 Using needle-nosed pliers, bend the ¹⁄₁₆" brass rod (connecting rod) to shape. See *both* the Parts View and Propeller Shaft detail for reference. Note that the bend on one end is 90° from that on the other end. Both ends need bending.

4 Working inside the crank opening, slip the crank onto the end of the connecting rod, then slide the crank onto the crankshaft where shown on the Section View drawing. Apply flux. Next, use a soldering gun to solder the crank onto the crankshaft (threaded brass rod) where shown on the Section View drawing, being careful not to solder the crank to the crankshaft tube.

5 Slide the top end of the connecting rod (attached to the crank in the previous step), into the hole in the pilot's scarf. Now, insert the pilot/scarf assembly into the hole in the cockpit.

6 Attach the propeller to the front end of the threaded rod. Adjust as necessary for smooth movement.

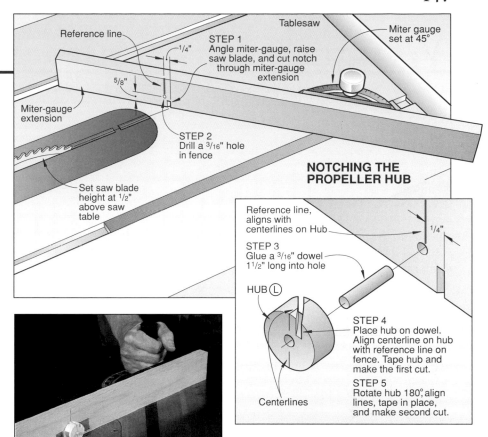

Shorten the rod if needed. Remove the propeller assembly.

FINISHING TOUCHES BEFORE TAKEOFF

1 Epoxy a ½"-long piece of ¼" brass rod into the end of a 4"-long piece of ⁹⁄₃₂" (I.D.) brass tubing (pivot tube) where shown on the Section View drawing. Now, epoxy the brass pivot tube into the hole in the bottom of the fuselage. Grind or file one end of a piece of ¼" brass rod 7" long (pivot pin) to the shape shown on the Section View drawing.

2 Finish-sand the wood parts. Mask the protruding ends of the brass tubing and threaded rod.

3 Apply a clear exterior finish to the propeller (we used spar varnish), and paint the propeller hub with an exterior enamel.

4 Apply two coats of aerosol primer to the plane. Follow up with two coats of red paint (we used Chinese red aerosol).

5 Detail-paint the pilot and scarf.

6 To make the emblem designs, either paint them onto the wings and fuselage, or transfer the patterns to colored tape. Apply the tape "decals" where shown on the opening photo.

7 Drill a ¼" hole in the top end of a post, and insert the ¼×7" brass pivot pin into the hole. Slide the whirligig onto the pivot pin.

BUYING GUIDE

• **Whirligig hardware kit.** ¹⁄₁₆" brass rod 6" long (connecting rod and pilot pivot pin), ¹⁄₁₆×½×4" solid brass (crank), #10-32 brass threaded rod 6" long (crankshaft), ⁷⁄₃₂" (I.D.) brass tubing 4¹³⁄₁₆" long (crankshaft tube), ⁹⁄₃₂" (I.D.) brass tubing 4" long (pivot tube), ¼"-dia. brass rod 8" long (pivot pin and pivot tube plug), 2–#10 brass nuts, 3–#10 brass flat washers, #18×⅝" brads. For current price of kit no. AP1, contact Miller Hardware, 1300 M. L. King Parkway, Des Moines, IA 50314. Or call 515/283-1724 to order. No CODs please.

Wetlands Silhouette

A simply stylish clock with a scrollsawed face

You can wade right into this project without any fear of getting in over your head. Our heron clock goes together quickly and simply, and it looks great standing on a shelf or hanging on the wall.

Cut parts A and B 1" longer than the sizes shown in the Bill of Materials. On each inside face, ¾" from the front edge, saw or rout a ¼" groove ⅜" deep for part C.

Miter-cut parts A and B to length, keeping the groove to the inside.

Cut part C to finished size. Then assemble parts A and B around C, glue, and clamp.

Cut part D to size. Draw on it a vertical centerline and a line perpendicular to the centerline 3¼" from the bottom. Scribe a 5"-diameter circle (2½" radius) centered on the lines' intersection.

Now, lay out 12 equally spaced points around the circle for the hour markers. Here's one way to do it: Without changing your compass setting, step off a series of six arcs around the circumference of the

circle, starting where the vertical line intersects the circle. Then, step off six more arcs starting from the horizontal line's intersection with the circle.

Photocopy the full-sized number patterns (you'll need five ones, two twos, and two sixes). Lay out the numbers inside the circle, keeping them vertical. Space the numbers visually, adjusting them for a pleasing appearance. Then, adhere them with rubber cement.

Photocopy the full-sized heron silhouette pattern. Fasten it to the stock, placing the top of the head at

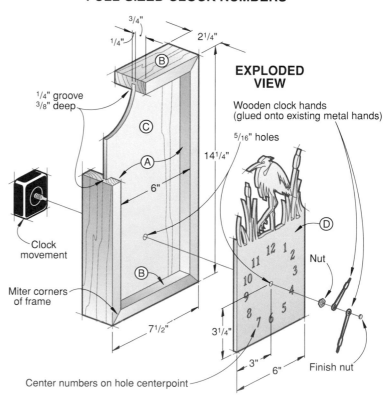

1234567890

FULL-SIZED CLOCK NUMBERS

EXPLODED VIEW

¾"

2¼"

¼"

B

¼" groove
⅜" deep

C

A

14¼"

6"

Wooden clock hands
(glued onto existing metal hands)

5/16" holes

D

Clock
movement

B

11 12 1
10   2
9   3
8   4
7 6 5

Nut

Miter corners
of frame

Finish nut

7½"

3¼"

3"

6"

Center numbers on hole centerpoint

the top edge of the wood. Drill blade start holes, and scrollsaw the numbers. Saw the interior parts first on those numbers that have them. Then cut out the silhouette, beginning with the small inside cuts. (We used a no. 4 blade, .033×.014" with 15 teeth per inch for all cutting.)

Sand the assembled frame and front. Glue the cut-out face into place on the front. Spear the small center pieces for the 4, 6, 8, 9, and 0 on the tip of an X-acto knife to glue them into place. Cyanoacrylate (CA) adhesive works great for attaching those small inside pieces.

Snip the teardrop-shaped ends from the clock hands. Cut the cattail hands out of veneer that contrasts with the clock face, and attach to the metal hands with epoxy or CA glue.

Drill a $^5/_{16}$" hole at the center of the clock circle. Sand the clock face as necessary, and apply a clear finish overall. When the glue has dried, insert the clock movement, secure with the nut provided, and install the hands.

BUYING GUIDE

• **Clock parts.** For current price of movement with hands and battery, product no. 71220, contact Klockit, P.O. Box 636, Lake Geneva, WI 53147, or call 800/556-2548.

Bill of Materials

Part	Finished Size*			Mat.	Qty.
	T	W	L		
A* sides	¾"	2¼"	14¼"	B	2
B* ends	¾"	2¼"	7½"	B	2
C front	¼"	6¾"	13½"	BP	1
D face	¼"	6"	11⅜"	W	1

*Start with longer stock, and cut to finished length in accordance with how-to instructions.

Materials Key: W-walnut, B-birch, BP- birch plywood

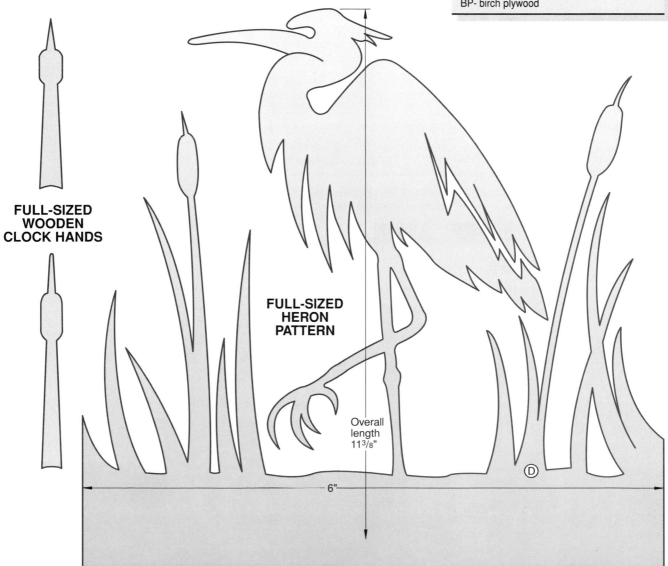

FULL-SIZED WOODEN CLOCK HANDS

FULL-SIZED HERON PATTERN

Overall length 11³/₈"

6"

North Woods Nester

Carve this waterfowl favorite

Here's classic waterfowl art you can carve—a loon gliding serenely on the water. Captured perfectly by Wisconsin woodcarver Rick Beyer, this loon delivers one-hundred percent pure carving enjoyment, not to mention beauty.

Enlarge the patterns on the *opposite page* at 200 percent. (The carved loon will be about ⅔ life size.) Transfer the enlarged patterns to your stock. Bandsaw the side and top views of the head and body blanks. Follow the outer pattern lines except at the back of the body on the side view; there, follow the dotted cutting line.

Draw a lengthwise centerline on top of the body. Parallel to the centerline and 1½" on either side of it, add two more lines. On each edge of the blank, draw a line parallel to the bottom and 1⅜" above it. Then, sketch in the neck location where shown by the Body Top View pattern.

Bevel the top of the body blank by cutting away the wedge between the outer top line and the edge line. Tilt your bandsaw table about 45° to do the job.

In the same fashion, bevel the square corners of the head (but not the beak). Cut to lines about ½" on either side of the centerline and about ½" from the top.

With a rasp or a rotary power-carver fitted with a toothed carbide cutter (Kutzall type), round the sharp corners and smooth the sawed surfaces of the head and body blanks. Shape the beak. Hollow out the back of the neck location—the part on the upward curve of the body—to match the bottom of the neck. At the front of the body, blend the contours to match the breast to the neck, as shown in Photo A.

SHAPE A LOVELY LOON
Sketch the outlines for the wing feather groups (the red pattern lines) onto the body. Then, add the blue-outlined side pockets.

A

Guidelines show feather groups for carving. When fitting the head, slope the flat top of the head to match the angle at the rear of the body.

Now, carve the feather groups. Take care as you shape the areas with a toothed carbide burr or fluted carbide cutter—you're establishing the final body contours. Don't try to make the edges of the groups look feathery, just establish the shape for each feature now. Later, you'll add detail to represent the feathers.

Refer to the photographs and the cross-section drawing as you work. The wings, folded over the body, form distinct humps on the bird's back. The body itself curves smoothly from behind the neck up to the high point on the back (which is at the wide point on the body), then down to the tail. Establish that line along the center of the loon's back.

continued

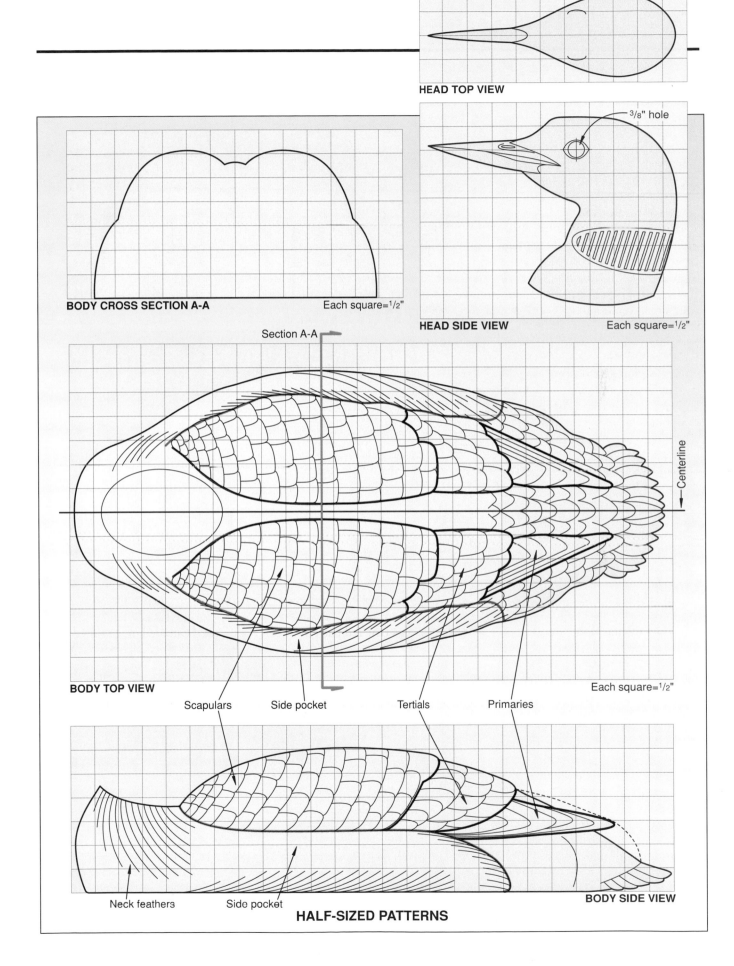

HEAD TOP VIEW

³/₈" hole

HEAD SIDE VIEW Each square=¹/₂"

BODY CROSS SECTION A-A Each square=¹/₂"

Section A-A

Centerline

BODY TOP VIEW Each square=¹/₂"

Scapulars Side pocket Tertials Primaries

Neck feathers Side pocket **BODY SIDE VIEW**

HALF-SIZED PATTERNS

North Woods Nester

continued

As you carve, remember that the wing groups overlap. The *scapulars,* the feathers nearest the body at the front of the wing, overlap the next layer back, the *tertials.* The *primaries,* the long feathers at the outer end of the wing, emerge from beneath the tertials at the back of the body.

The primaries don't lie flat against the body at the back. So, with a tapered cutter, remove stock beneath the ends of the primaries. (Later you'll undercut them more.) Make the primaries straight rather than curving them to match the body contour.

Part of the wing tucks into the side pocket, so model the upper edge of the pocket overlapping the scapulars and tertials. The pocket also opens at the back. Carve a slight overlap where the back end lies against the body.

Smooth the body with 100-grit sandpaper after laying out the feather groups. Soften the breaks between the groups, but don't sand so much that you eliminate those separations.

GIVE YOUR LOON A NECKLACE

Drill a ⅜" hole about ⅜" deep at the eye location on each side of the head. Draw the bullet-shaped necklace-feather group on each side of the neck. At the back of the neck, leave about ⅜" between the left-side and right-side groups.

Carve around the outline with an inverted-cone burr, as shown in Photo B. After defining the feather group, carve away the neck above and below it to raise the area a little more than ¹⁄₃₂" above the surrounding surface.

Draw the necklace feathers on the raised area. Then, with a small flame-point ruby carver, cut in to separate and shape the feathers as shown in Photo C.

Cut all the way to the neck surface—but no deeper—around and between the feathers. The only trace that should remain of the raised area you carved earlier is the necklace of feathers.

PUT ON THE PLUMAGE

Starting from the front of the body, pencil in the feathers, as shown in Photo D, *opposite.* Refer to the photographs and patterns. (You may want to gather some additional book or magazine photos of loons, too.)

Sketch in the neck feathers where shown, curving them downward and back on each side of the body. Breast feathers flow down the front. You can resolve a lot of feathering questions by imagining the loon facing directly into a strong wind.

With the inverted-cone cutter, shape the tail feathers as shown in Photo E, *opposite.* Note that the feathers slant from the left edge down to the right edge on the left side of the tail, the opposite way on the right side.

Carve the body feathers with fluted burrs and ruby carvers. Separate the feathers within the wing groups as shown in Photo F, *opposite.* Model the tips of the side-pocket feathers as shown in Photo G, *opposite.* Don't carve too deeply—you're representing feathers, not shingles.

Refer to the patterns and photos to carve the primary feathers. Then, undercut the primaries with a knife. Undercut to a depth of 1⅜" at the outside, all the way to the tertials on the inside. The thickness at the tip should be about ⅛".

NOW, TEND TO SOME DETAILS

Mix a small amount of two-part epoxy putty. Fill the eye hole on one side with putty, then press in a 10 mm red glass eye. Install the other eye, and check them for symmetry as

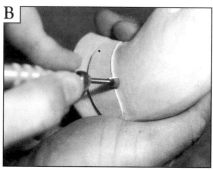

Raise the necklace-feather area with an inverted cone burr.

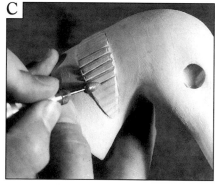

Separate the necklace feathers first, then shape the ends. The finished necklace will stand above neck surface ¹⁄₃₂" or less.

shown in Head (Front View) drawing, *opposite.*

Sculpt the eyelids with a knife, referring to the photographs. Cut the elongated nostrils into the upper mandible with a knife or small rotary bit.

After delineating the feathers, establish the layered look by grinding down the front of each segment slightly with a ruby carver.

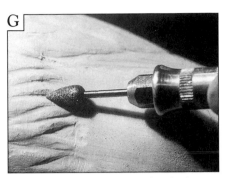

Instead of modeling each individual feather on the breast and sidepockets, suggest them by carving soft, flame-like streamers.

You don't need to draw every feather, just sketch in the general lay of the plumage.

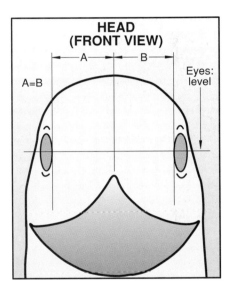

HEAD (FRONT VIEW)

A=B

Eyes: level

You can now paint the basic loon in a black-and-white scheme, following the photographs. Or, for a more detailed carving, refer to Rick's comprehensive detailing and painting instructions (see the Buying Guide).

TOOLS AND SUPPLIES

- **Stock**
  Basswood or other carving wood; 2×5½×4" for the head, 3½×6×13" for the body (see Buying Guide below for bandsawed blanks)
- **Power carving equipment**
  Flexible-shaft rotary carving machine or rotary hand tool
- **Bits**
  Carbide-tooth tapered burr; ½ or ¾" (Kutzall silver T-12 or T-34)
  Fluted carbide burr; ¼" inverted cone, ¼" or larger taper
  Ruby carver; flame point, small and large
- **Knife**
  Bench knife or X-Acto knife

Rough in the tail feathers about ⅛" thick. Finished feathers display a convex upper surface, a flat or slightly concave bottom.

Attach the head to the body with woodworker's glue and a dowel. After the glue dries, grind a V-groove around the neck-body joint, and fill it with two-part epoxy putty. Sand the joint, then sand the entire loon as necessary.

BUYING GUIDE

- **Painting guide.** For current price of detailed painting instructions, contact R.J. Beyer Galleries, 1115 N. Main St., Racine, WI 53402, 414/633-7172.

Signature Shop Clock

A tradition-rich timepiece that says it all

We designed our Signature Shop Clock for Idea Shop™ 2, our testing ground of innovative shop ideas. Like the other hardworking projects in Idea Shop 2, this wall-mounted clock had to meet two criteria. First, it had to perform its task well. And second, it had to be an attractive addition to our workshop. To address the first, we selected a clockface with large numerals so that you can tell time easily from anywhere in your shop. And as for the second criteria, we think you'll agree this clock is good enough to grace the wall of any workshop.

To personalize it, order your own computer-engraved plate. See the Buying Guide for ordering details.

THE BACKBOARD COMES FIRST

1 Edge-join enough ¾"-thick walnut to form a 13" wide by 14" long backboard blank.

2 Trim the backboard (A) to the size noted on the Backboard drawing. Transfer the clock-opening centerpoint and the mounting-screw centerpoints to the blank. Cut the backboard to shape. Sand the edges to remove the saw marks.

3 Working from the clock-opening centerpoint, cut a 3⅜"-diameter hole in the backboard. (We drilled a blade start hole and scrollsawed the hole to shape. You also could use a circle cutter to form the opening.)

4 Using the two mounting-screw centerpoints you marked earlier, drill a pair of ⁵⁄₃₂" countersunk holes through the backboard. Don't worry; the clockface will hide the screws later.

5 Rout a ¼" classic cove along the front edge of the backboard. See the Edge Detail accompanying the Exploded View drawing for

continued

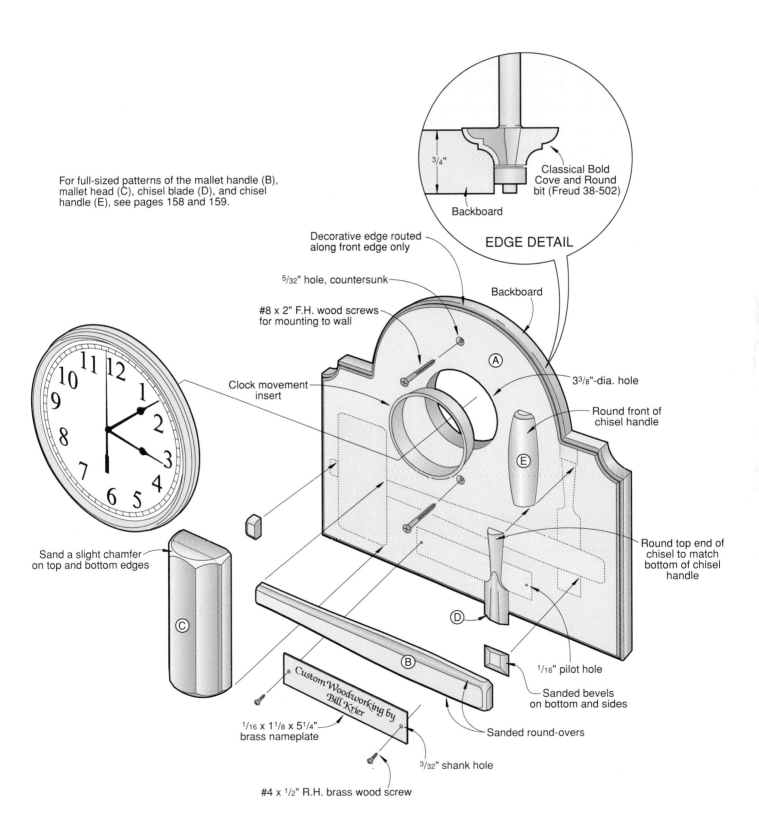

For full-sized patterns of the mallet handle (B), mallet head (C), chisel blade (D), and chisel handle (E), see pages 158 and 159.

Classical Bold Cove and Round bit (Freud 38-502)

Backboard

3/4"

EDGE DETAIL

Decorative edge routed along front edge only

Backboard

5/32" hole, countersunk

#8 x 2" F.H. wood screws for mounting to wall

Ⓐ

3³/₈"-dia. hole

Round front of chisel handle

Clock movement insert

Ⓔ

Round top end of chisel to match bottom of chisel handle

Sand a slight chamfer on top and bottom edges

Ⓒ

1/16" pilot hole

Ⓓ

Sanded bevels on bottom and sides

Ⓑ

Custom Woodworking by Bill Krier

1/16 x 1¹/₈ x 5¹/₄" brass nameplate

3/32" shank hole

Sanded round-overs

#4 x 1/2" R.H. brass wood screw

EXPLODED VIEW

Signature Shop Clock

continued

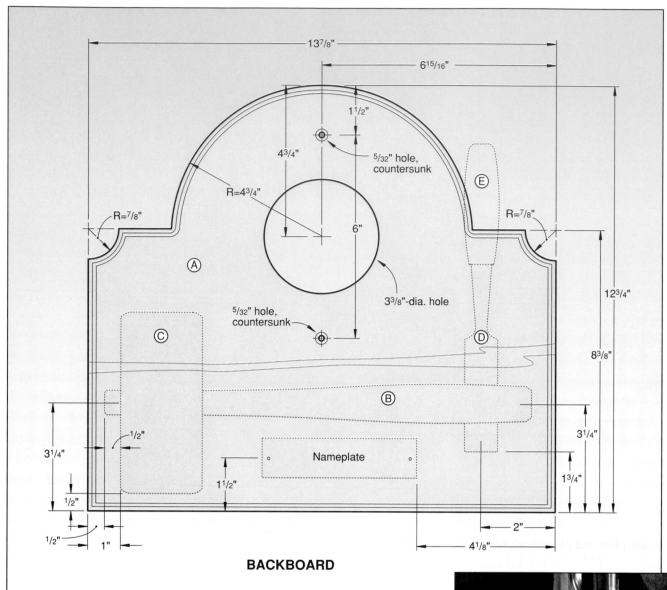

13⁷/₈"

6¹⁵/₁₆"

1¹/₂"

4³/₄"

⁵/₃₂" hole,
countersunk

R=4³/₄"

R=⁷/₈"

R=⁷/₈"

E

6"

12³/₄"

A

3³/₈"-dia. hole

⁵/₃₂" hole,
countersunk

C

D

8³/₈"

B

3¹/₄"

1/2"

1³/₄"

3¹/₄"

Nameplate

1¹/₂"

1/2"

2"

1/2"

1"

4¹/₈"

BACKBOARD

reference. (We used a Freud classical bold cove and round bit, #38-502.) Finish-sand the backboard.

OK, LET'S FASHION THE MALLET

1 To make the mallet handle (B), cut a piece of ¾" ash to 1¼×10⅜". Make a photocopy of the full-sized Front and Side View patterns on *page 158*, and attach them to the handle blank with spray adhesive.

2 Bandsaw the handle Side View to shape the top of the handle. Tape

the waste stock back in place (it's got the Front View pattern attached), and then cut the Front View pattern to shape as shown in the photo *at right*.

3 Sand round-overs along the top edges of the handle where shown on the Section View portion of the full-sized Front View handle pattern on *page 158*. Crosscut a ½"-long piece off the end of the mallet handle, and set both pieces aside for now.

After cutting the handle side view to shape, tape the waste back in place, and cut the front view pattern to shape.

Bevel-rip the edges of the laminated-ash mallet-head blank at 45°.

4 To form the mallet head (C), glue and clamp two pieces of ³⁄₄×3×10" ash face-to-face. (We made the mallet-head blank 10" long for safety when bevel-ripping it to shape.) Rip and plane both edges for a 2³⁄₈" finished width, and then plane or resaw for a 1¹⁄₈" final thickness.

5 Bevel-rip the edges at 45° where shown on the Mallet Head Full-Sized Patterns drawing on *page 159*. See the photo *above* for reference.

6 Crosscut a section 5³⁄₈" long from the 10"-long mallet-head blank. Then, sand a slight chamfer (we used a stationary disc sander and palm sander) around the ends of the mallet head.

7 Finish-sand the parts. Next, using the Exploded View and Backboard drawings for reference, glue the mallet parts (B, C) to the

backboard (A). Don't forget to add the ¹⁄₂"-long mallet handle end piece.

ONE SHARP-LOOKING CHISEL COMING UP

1 Cut a piece of ¹⁄₂" maple to 1×4¹⁄₂" for the chisel blade (D). Make a photocopy of the Chisel Blade Side and Front View patterns on *page 159* and adhere them to the blank. Cut the Side View pattern to shape. Use double-faced tape to adhere the waste stock (it's got the Front View pattern on it) to the chisel blank.

2 Using the Front View pattern and Blade Section views on *page 159* as a reference, cut and sand the blade top surface and bevels to shape. (You might find it just as easy to use your own 1"-wide chisel as a guide when shaping the blade.)

3 Crosscut the chisel blade into two pieces where shown on the Chisel Blade Full-Sized Patterns drawing. To ensure the glued-down chisel blade pieces appear straight (the portion of the mallet handle they're glued next to is curved), contour-sand the cut edges of the blade to butt snugly against the curved mallet handle.

4 Cut and sand the chisel handle (E) to shape from ³⁄₄"-thick cherry.

5 Finish-sand all the pieces, and glue them to the walnut backboard where shown on the Backboard drawing.

ADD THE FINISH, AND HANG UP THE CLOCK

1 Apply the finish of your choice. (We sprayed on three coats of aerosol polyurethane; aerosol lacquer also would work.) Position and screw the brass plate in place. (See the Buying Guide for our source of custom-engraved brass plates.)

2 Screw the clock to the wall with a pair of #8×2" flathead wood screws or toggle bolts. Install the battery in the clock, and fit the movement into the hole in the backboard.

BUYING GUIDE
• **Quartz clock insert.** 8¹⁄₄"-diameter insert with Arabic dial. Stock no. 15141. For current prices contact Klockit, P.O. Box 636, Lake Geneva, WI 53147. Or call 1-800-556-2548 to order.

• **Engraved plate.** 1¹⁄₈×5¹⁄₄" satin brass with the words "Custom Woodworking by (your name)" in black double-lined letters. Two screws included. For current prices contact Custom Awards, 1425 22 Street, West Des Moines, IA 50266. No phone orders please.

continued

Signature Shop Clock

continued

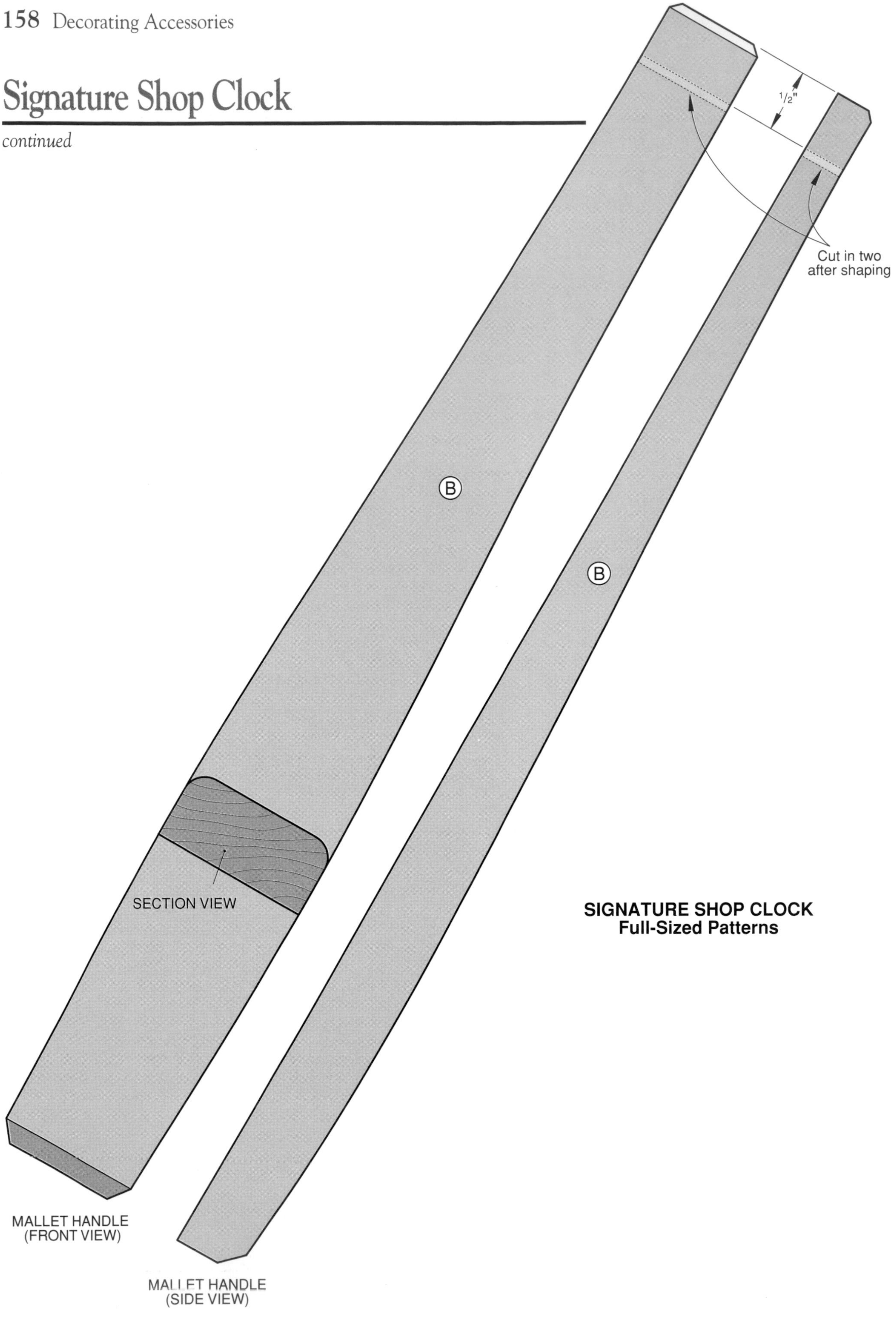

1/2"

Cut in two
after shaping

(B)

(B)

SECTION VIEW

SIGNATURE SHOP CLOCK
Full-Sized Patterns

MALLET HANDLE
(FRONT VIEW)

MALLET HANDLE
(SIDE VIEW)

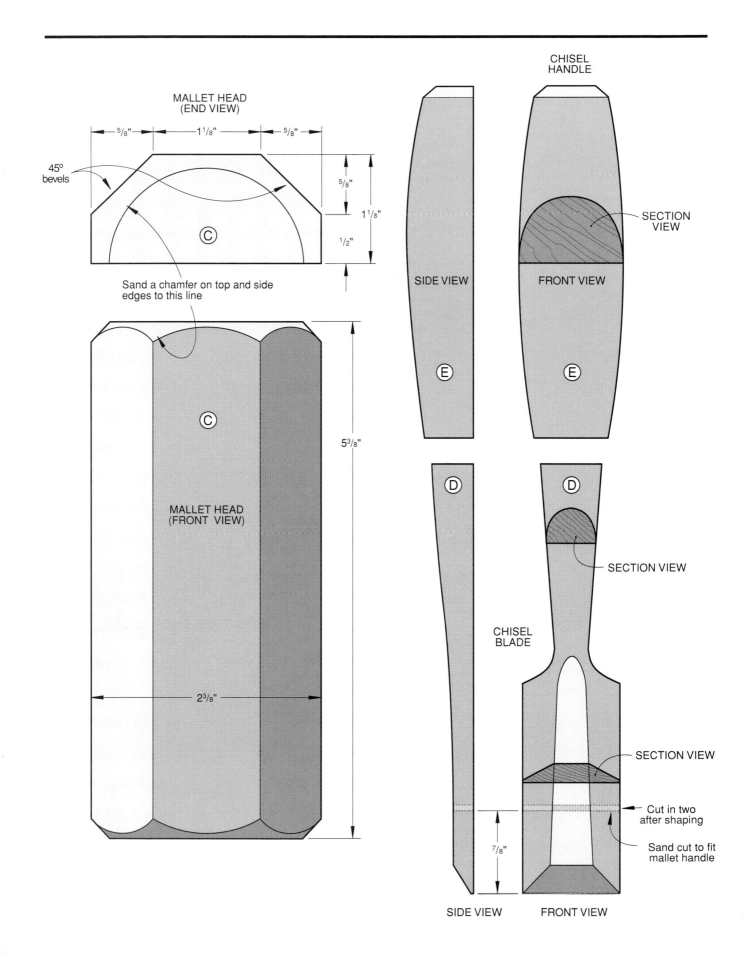

MALLET HEAD
(END VIEW)

5/8" 1 1/8" 5/8"

45°
bevels

5/8"

1 1/8"

1/2"

C

Sand a chamfer on top and side
edges to this line

MALLET HEAD
(FRONT VIEW)

C

5 3/8"

2 3/8"

CHISEL
HANDLE

SIDE VIEW

FRONT VIEW

SECTION
VIEW

E

E

D

D

SECTION VIEW

CHISEL
BLADE

SECTION VIEW

7/8"

Cut in two
after shaping

Sand cut to fit
mallet handle

SIDE VIEW   FRONT VIEW

Acknowledgments

WRITERS

Written by Larry Johnston—A Reel Deal for Kids, pages 73-75; Scrollsawed Safari Puzzle, pages 82-83; Cupid's Mystery Valentine, pages 102-103; Restaurant in the Round, pages 104-107; Turned Perfume Applicator, pages 108-109; A Box That's a Bit Fishy, pages 110-111; Block Prints, pages 112-115; Saint Nick, pages 124-125; Tour-De-Force Tureen, pages 126-130; Wetlands Silhouette, pages 148-149; North Woods Nester, pages 150-153

Written by Marlen Kemmet—The Do-It-All Mobile Tablesaw Base, pages 7-11; Accomodating Cabinets, pages 12-15; Drop-Leaf Mobile Workbench, pages 16-22; Angle-Master Miter Jig, pages 23-28; Finishing Center, pages 29-32; Comfy Country Chair, pages 34-37; Picnic-Perfect Table and Benches, pages 38-41; Terrific Tambour Bookcase, pages 42-47; Masterpiece in Pine, pages 48-53; Stylish Storage, pages 54-59; Show-Off Showcase, pages 60–63; Corner Classic, pages 64–71; Sunny-Day Sandbox, pages 76-79; Victory-Lap Clothes Rack, pages 80-81; Pickup from the Past, pages 84-87; Snow Stroller, pages 88–92; The Fun-Time Racer, pages 93-100; Tabletop Reindeer, pages 116-117; Dresser-Top Delight, pages 118–123; Show-Stopping Picture Frames, pages 132-135; Wetlands Visitors, pages 136–141; Ace-of-Hearts Triplane Whirligig, pages 142–147; Signature Shop Clock, pages 154-157

PROJECT DESIGNERS

Rick Beyer—North Woods Nester, pages 150–153

Jim Boelling—Stylish Storage, pages 54–59

Tom Brahill—Wetlands Visitors, pages 136–141

James R. Downing and Jim Boelling—Angle-Master Miter Jig, pages 23–28

James R. Downing—The Do-It-All Mobile Tablesaw Base, pages 7–11; Accomodating Cabinets, pages 12-15; Finishing Center, pages 29-32; Comfy Country Chair, pages 34-37; Masterpiece in Pine, pages 48-53; Corner Classic, pages 64-71; Sunny-Day Sandbox, pages 76-79; Dresser-Top Delight, pages 118-123; Show-Stopping Picture Frames, pages 132-135; Ace-of-Hearts Triplane Whirligig, pages 142-147; Signature Shop Clock, pages 154-157

Jamie Downing—Victory-Lap Clothes Rack, pages 80-81

Glenn Crider—© A Reel Deal for Kids, pages 73-75

C.L. Gatzke—Tour-De-Force Tureen, pages 126-130

George Hans—Tabletop Reindeer, pages 116-117

Peggy Johnston—© Scrollsawed Safari Puzzle, pages 82-83; Block Prints, pages 112-115

Parman Brothers LTD., Johnson, Kansas—Terrific Tambour Bookcase, pages 42–47

Ron Pavelka—Restaurant in the Round, pages 104–107

Peach Crafts, Gary Peach—Turned Perfume Applicator, pages 108–109

Erv Roberts—Drop-Leaf Mobile Workbench, pages 16–22

Bill Trumble—Pickup from the Past, pages 84–87

Fern Weber—© Saint Nick, pages 124–125

Thomas J. Wiley, Pied Piper Productions—A Box That's a Bit Fishy, pages 110–111

Bill Zaun—© Wetlands Silhouette, pages 148–149

PHOTOGRAPHERS

King Au
John Hetherington, Hetherington Photography
Bill Hopkins
Wm. Hopkins, Wm. Hopkins Photography
Harvey Nyien

ILLUSTRATORS

Jamie Downing
Kim Downing
Brian Jensen
Roxanne LeMoine

Enlarging gridded pattens by hand

Gridded patterns in this book that require enlargement include the statement "Each square = 1"." This means that grid squares in the drawing *must* be enlarged to the size indicated for your full-sized pattern.

To use the hand-enlargement method called transposing, you'll need cross-section graph paper (the kind with heavier lines marking off each square inch), a ruler, an eraser, and a soft-lead pencil. (If graph paper isn't available, make your own by dividing plain paper into the specified-size squares.)

Begin by marking off on your grid paper the same number of squares as indicated on the pattern grid. Next, number each vertical and horizontal grid line in the pattern. Then, number the corresponding grid lines on your graph paper the same way.

Start your pattern enlargement by finding a square on your graph paper that matches the same square on the original gridded pattern. Mark the graph paper grid square with a pencil dot in the same comparative place where a design line intersects a grid line on the original. Work only one square at a time. Continue to neighboring squares, marking each in the same way where a design line intersects a grid line.

To avoid discovering any mistakes too late, mark only part of the design, then stop and join the dots with a pencil line. For more precision, draw all of the straight lines first; then add the curved and angled lines. Once you have transposed part of the design, finish marking the rest of the squares and join those dots in the same way.

Sometimes, you'll only have a *half-pattern* to use. To duplicate a full-sized half-pattern, copy the original with a soft-lead pencil on tracing paper. Next, flip your traced pattern over and place it pencil-lines-down onto one half of the board. After aligning the pattern for position, go over the pattern lines with your pencil to imprint it on the board. Then, flop the pattern onto the second half of the board and again retrace the pattern to imprint it. This method proves faster than copying with carbon paper and doesn't mark up the original pattern.

Using a copier to enlarge a gridded pattern

A photocopier with enlargement capability enlarges a pattern faster than transposing. (Even a copier can be a little inaccurate, so always check your results with a ruler.)

To find out the enlargement percentage you'll need, use a pocket calcuator to divide the scale square size (1") by the actual size of a gridded pattern square (for example, ½"). Your resulting enlargement will need to be 200% of the original.

However, the copier you use may only have an enlargement limit of 150%. If this is the case, make a first enlargement of the original at 150%. Next, divide your desired final enlargement percentage (200) by 150. Your answer will be 133.

Then set the photocopier at 133%, and make a second enlargement of your first copy (which was made at 150%), and you'll end up with a pattern that is 200% larger than the gridded pattern. Check the final copy with a ruler to ensure it is sized correctly. If the final copy isn't exactly correct, adjust the copier up or down a percentage or two until you end up with a pattern that is the correct size.

U.S. Units to Metric Equivalents

To convert from	Multiply by	To get
Inches	25.4	Millimeters (mm)
Inches	2.54	Centimeters (cm)
Feet	30.48	Centimeters (cm)
Feet	0.3048	Meters (m)
Yards	0.9144	Meters (m)

Metric Units to U.S. Equivalents

To convert from	Multiply by	To get
Millimeters	0.0394	Inches
Centimeters	0.3937	Inches
Centimeters	0.0328	Feet
Meters	3.2808	Feet
Meters	1.0936	Yards

If you would like to order any additional copies of our books, call 1-800-678-2803 or check with your local bookstore.